POWER PROFITS

A Comprehensive 9-Step Framework For Reducing Electricity Costs and Boosting Profits

Michael Williams

First published in 2018 by Michael Williams

A catalogue entry for this book is available from the National Library of Australia.

ISBN: 978-0-9954461-0-6

Printed in Australia by McPherson's Printing
Project management and text design by Michael Hanrahan Publishing
Cover design by Julia Kuris

The paper this book is printed on is certified as environmentally friendly.

Contents

Part III: Applying the framework

Introduction

The energy landscape is changing – rapidly

One of Australia's international key competitive advantages over many decades has been access to reliable and relatively cheap energy, whether it be electricity, gas or coal. We have structured our manufacturing equipment and infrastructure around this cheap energy.

When energy costs have been low it's been easy to overlook the inefficiencies in our power use. But the energy landscape has changed suddenly and rapidly, and today energy costs are increasing. This means that – suddenly – businesses and households are starting to pay a lot more attention to both how they use energy and where and how they obtain it.

The changes started gradually with the development of regulated national electricity and gas markets and the subsequent privatisation of electricity and gas networks and electricity generation starting in the 1990s. It gathered momentum in recent years with market incentives to drive growth in renewable energy such as wind and solar, disrupting the traditional models. It escalated with the internationalisation of our natural gas resources through development of massive natural gas liquefaction and export facilities in Queensland and Western Australia. It has been made worse by the inevitable closure of traditional base-load, old and inefficient coal-fired power stations.

In 2016 Eastern Australia (excludes WA and NT) produced approximately 1.6 petajoules (PJ) of gas but only 0.6 PJ was consumed in the domestic market; 1 PJ of natural gas was compressed, liquefied and exported to our major trading competitors. In 2014 there was next to no Eastern Australian Liquefied Natural Gas (LNG) production. Much of the natural gas that supplies our households, small businesses, and manufacturing and industrial processes is sourced from the same resource as the gas that is exported. It's true that much of these gas resources and subsequent economic benefits in terms of jobs, growth and royalties would not have been realised without an LNG export market, but the tradeoff is the internationalisation of one of our very last manufacturing competitive advantages.

While the giant LNG plants were built on the basis of meeting their feedstock needs from so called "unconventional" coal seam methane gas resources that had not been previously developed, in reality the "conventional" gas sources such as the Cooper Basin resource have had to supplement the supply of natural gas to these plants. This conventional gas had previously been used exclusively in the domestic market for households and industry. As a result, demand for conventional natural gas increased significantly and so did prices. Prices rose to the level that the LNG plants were prepared to pay. That price, in simple terms, was the international traded gas price less liquefaction costs at the LNG plants.

After decades of very low gas prices compared with our international competitors, we began to pay a price that reflected the international price. In the space of two to three years, natural gas prices rose by more than 100%.

With rising gas prices comes rising electricity prices, as often gas-fired generators are setting the half-hour wholesale price for electricity. It also means that gas-fired generation is less competitive with coal-fired generation and coal-fired generators can safely increase their bid prices in the "competitive" wholesale market. The two markets, gas and electricity, are inexorably linked. The outcome is much higher electricity prices for households, small businesses and the commercial and industrial sectors.

There are other factors pushing up electricity prices such as increases in transmission and distribution supply charges, the closure of old and inefficient coal-fired power stations and, to

some extent, environmental charges, but the KO punch has been delivered by internationalising our domestic gas resources.

So why have I written this book?

I am passionate about helping Australian manufacturing businesses increase their international competitiveness and making Australia a great manufacturing nation to provide jobs and economic security for our kids.

As an "operations" professional I have always been passionate about increasing efficiency, reducing costs and increasing profits. In 2001 I left BHP after a 19-year career as a production metallurgist and line manager at some of the most energy-intensive businesses in Australia, including the Whyalla Steelworks, The Port Kembla Steelworks, the Tasmanian Electro Metallurgical Company (TEMCO) and the Hot Briquetted Iron Project in Port Hedland. These businesses all had electricity demand in the hundreds of megawatts (MW) and energy bills in the hundreds of millions of dollars per year. That has a much bigger impact on the bottom line than office supplies. Energy costs were so critical to these businesses that – even in an era of low energy costs – they invested significantly in their own electricity generation and energy recovery. The electricity use was so high that control of maximum demand was critical to ensure that they did not incur very expensive network supply costs. Energy was so critical that the primary key performance indicator was electricity used per tonne of product, or kilowatt-hours per tonne (kWh/t).

My early days in production at these very energy-intensive operations led me to develop a high awareness of energy efficiency and reducing energy costs. As the energy intensity was so high, usually any investment initiative to improve energy efficiency had a very quick payback and was generally supported and implemented.

Roll forward to 2001 when I joined the cement industry, which is also a large energy user but not as large as the very energy-intensive steel and metal smelting and refining industries. Energy had always been important and was in the top two input costs, accounting for 20% to 25% of total costs, with energy being used in the form of electricity to power large drives, and coal and gas to provide process fuel for heating. However, because energy use was not of the same order of magnitude as the large metal smelters and refineries, there had not been the same level of focus on energy efficiency, energy recovery, electricity self-generation, and control of maximum demand. It was still important but it was a secondary key performance indicator. The primary metric was always tonnes produced.

Investments to improve energy efficiency had to be very compelling to be approved, and usually required very quick payback periods of less than two years. Investments in new plant equipment to expand production focused on minimising the capital cost of the project, not maximising the energy efficiency of the new equipment. As with other similar industries, the lack of investment in energy efficiency over the last two decades has left those businesses much more sensitive to increases in electricity costs than if they had invested in more efficient equipment.

This book is written for business owners or managers who are concerned, or even alarmed, at the rapid increase in energy costs, particularly electricity prices, and who would like to take control of their own destiny rather than be a victim of increasing energy costs.

The book is written in the context of the Australian East Coast National Electricity Market (NEM), with a distinctive South Australian flavour, as the best flavours are found in South Australia, but the principles can be applied in any electricity market in the world, and even in the special state of Western Australia. And the 9-Step Electricity Cost Reduction Framework that we outline in this book looks to reduce your costs and increase business profits by reducing your electricity consumption, reducing your electricity unit prices, and reducing your so-called fixed electricity supply charges.

I hope that you enjoy reading it, and apply some or all of the framework to reduce your electricity costs and boost profits.

PART I

Electricity as a strategic business issue

CHAPTER 1

Understanding electricity in Australia

The Australian economy has enjoyed decades of low energy costs in comparison with our international competition, in the form of cheap coal, cheap natural gas and cheap electricity. This would seem like a boon for both businesses and households. Companies have been able to get away with not paying too much attention to their energy consumption. Many organisations consider energy to be just another supply that they purchase, no different to office supplies, furniture or software – they aim to get the cheapest price, but there's no strategy involved in the purchase and use of electricity. And while many households have made some effort over the years to be careful with their energy use, for most people it was not a significant issue.

But there has also been a downside to this cheap energy. Our electricity costs have historically been so low that – to reduce capital costs – we have built plant, equipment and appliances that are inefficient, and we have been wasting an enormous amount of energy because it has been cheap.

Energy markets and the government

The question is often asked as to whether the government should do something to prevent the rise of energy costs. Both sides of politics are committed to free and efficient markets and are very reluctant to intervene. However, they can and do drive market reforms to make energy markets more efficient. One example is the recent initiative to regulate the gas market to promote trading of contracted but unused gas pipeline capacity to increase gas supply to market hubs. This facilitates the flow of gas to a gas-fired generator that may not have otherwise flowed as the generator may not have contracted for that capacity. This allows that generator to then compete for the supply of electricity and potentially puts downward pressure on prices.

It is no longer the role of government to ensure the supply of low fixed-priced electricity to end users. It is the role of government to establish, regulate and monitor energy markets that ensure the long-term security of supply at the lowest economical price.

Market reforms do help, but the fact remains that Australian gas and electricity costs are increasing due to the massive changes in the gas market, rapid growth in renewable electricity generation, and closures of old brown or sub-bituminous coal-fired generation plants such as the Northern Power Station at Port Augusta in South Australia (in 2016) and the Hazelwood Power Station in Victoria (in 2017).

Energy markets and businesses

To ensure that rising energy costs do not erode profits, Australian businesses must be prepared to innovate, invest and be flexible to achieve lower costs and even increase profits.

Comments I hear often from businesses include:

- "We can't afford the risk of price uncertainty – we need guaranteed fixed prices."
- "Electricity generation is not our core business, so we are not investing in our own generation."
- "We won't be investing in energy efficiency unless the payback is less than two years."
- "The government needs to do something about energy prices."
- "Our core business is not in supporting gas exploration and development."

- "The Renewable Energy Scheme is pushing prices up and should be abolished."

These are the responses of businesses that do not realise the competitive advantage to be gained from a well planned and executed energy strategy, and these businesses will be left behind as the energy market continues to change rapidly in coming years.

But there are many smart and agile businesses that *are* taking action now to not only protect themselves from rising costs but to actually reduce energy costs. For example, many businesses are:

- reducing input prices by gaining exposure to wholesale market prices rather than locking in high fixed prices
- investing in alternative generation to hedge against high market prices, or are even supplying their own electricity
- investing in energy-efficiency initiatives to drive down total energy use in the long term
- doing deals with gas developers to promote exploration and development, and are securing options on lower price gas in the future
- securing supply of waste-derived methane to use as a process fuel or to generate electricity
- utilising fuels derived from waste products such as construction wood waste, plastics, tyres, waste oil and waste solvents
- investing in biofuel projects
- investing in solar and wind generation to reduce net energy consumption, generate their own renewable energy certificates and earn additional revenue
- contracting with renewable energy projects for long-term offtake agreements at well below current wholesale prices
- flying beneath the radar with their energy plans to maintain their competitive advantage.

Businesses must innovate in how they manage their energy requirements or their profitability will suffer. A good energy strategy can reduce energy costs by as much as 50% compared with traditional thinking.

Energy strategies and capital investment

So how do large publicly owned manufacturing businesses make a decision on where they invest their capital?

Irrespective of the profitability and size of the business, the amount of capital that is invested in any year to sustain, de-risk or grow the business is limited, and the "shopping list" is usually much larger than the funds available. Some years the available capital may be very low or none at all. So growth, efficiency, equipment replacement, safety, quality, environmental and staff amenity projects are all fighting for a share of limited capital.

As a general rule of thumb in publicly owned manufacturing businesses, capital investment in new plant and equipment designed to increase efficiency and reduce costs usually requires a simple payback in less than two years. In tough times that payback period is often reduced to one year.

However, for strategic projects that grow the business, capital decisions may be based on a 10-year rate of return (IRR) or net present value (NPV) and not on a simple payback period. Stay in business (SIB) capital decisions for end-of-life equipment, or equipment in poor condition, are often based on risk. In this case, risk is the likelihood of a piece of equipment failing combined with the potential consequence of failure in terms of financial, safety, customer, environmental or reputational risk.

Energy costs are generally treated as simple input costs and not as a strategic business growth element. This means that energy-efficiency or cost-reduction measures are usually required to have a payback of one to two years, and so then they have to compete for funds against the other important initiatives.

Let's have a look at an example. Let's say we have a strict capital budget for the next financial year of $5 million. We have the following projects that are competing for capital:

1 $3m plant expansion that increases sales and profit with an internal rate of return of 25%.

2 $1m to replace a piece of equipment at end of life that has very high probability of failure and a consequence of $0.5m loss and disruption to customer supply. There is no financial payback.

3 $1m to install guarding around equipment to comply with safety legislation, with no financial payback.

4 $1m to invest in energy-efficiency projects that have a four-year simple payback and an IRR of 30%.

Guess which project misses out? The energy project, despite having the best rate of return.

Project 1 goes ahead as it has the largest impact on growing the business and growing revenues. Project 2 goes ahead as there is a high risk that there will be a failure that has a material impact on the business. Project 3 goes ahead to ensure the safety of employees, and as it is a legislative compliance issue. That leaves no capital left for the highly commercially viable energy-efficiency initiatives.

How has this played out in the cement industry? Despite high levels of waste heat from kilns and even cement mills, the payback for investment in waste heat recovery was historically always much greater than four years due to low electricity costs and the high costs of the investment required to recover that heat and generate electricity. The projects could have a high IRR but the payback period and the level of capital required were too large.

Electricity hasn't generally been seen as a strategic issue, rather just an input cost issue. Kiln fuel costs, on the other hand, were seen as a strategic issue. The industry had started to become very innovative in developing alternative fuels derived from waste to reduce fuel costs and reduce reportable carbon emissions. The individual cement companies saw this as a means of gaining a strategic competitive advantage over their competitors and to increase profits. Two of the three major cement companies in Australia invested early in businesses to specifically source and process waste-derived fuels for use as fuel in kilns, and the third company is trying to catch up. Investment payback in alternative fuels was often much longer than two years and had a high degree of risk around success. The projects were justified on the basis of a long-term rate of return. Those investments are now paying back significantly as coal and gas prices have increased.

Investments to reduce electricity costs

So what about investments to reduce electricity costs?

Deregulation of the various state electricity markets was adopted in Australia to attract private investment and reduce public funding, promote competition, provide choice to consumers and with the ultimate aim of reducing costs for households and businesses. This also facilitated privatisation of some or all of the electricity supply chain to promote efficiency and reduce state debt.

With progressive deregulation of the electricity market and the major electricity users becoming contestable customers in the late 1990s and early 2000s, a few innovative manufacturing companies refused to accept and lock in high-priced electricity retail supply contracts. They instead opted to take their chances being exposed to the potential volatility of the wholesale spot market.

These innovative companies were initially very worried about the lack of certainty in prices and the potential losses if high price events occurred. After developing and implementing risk management strategies to minimise exposure to high prices, they were very pleasantly surprised at how low their average annual prices were compared with the supposedly low-risk fixed retail prices.

More than 15 years later, many other medium to high electricity-using businesses joined the early pioneers to purchase electricity at wholesale prices. They are taking the risk of short-term high prices and price volatility in return for much lower annual average prices than they would have experienced had they signed up to fixed retail contracts.

As a direct result of being exposed to short-term prices they have come to better understand their electricity usage patterns and drivers, and have used this knowledge and information to achieve not only lower unit costs of electricity but also other efficiency cost savings and lower "fixed" network supply charges. The irony, though, was that the very successful wholesale price purchasing strategy kept electricity costs very low and so investment in electricity efficiency and waste heat recovery became much more difficult to justify as the payback periods were too long.

As electricity prices increase, the options to reduce costs through capital investment and innovative purchasing strategies also increase. The large increase in electricity wholesale prices in 2016 that has also flowed into retail prices has created greater incentives to invest in energy-efficiency initiatives. Initiatives that were shelved earlier this decade or even in the previous decade as having a low

IRR or long payback are now likely to be highly viable. Furthermore, if rising electricity costs really are a major concern to a business, they should treat electricity as a strategic issue and not just a utility cost input to put out to simple competitive tender every three years.

Those businesses that choose not to have a strategy for energy cost reduction, not to invest in efficiencies or alternative generation methods, not to analyse their electricity usage patterns to identify savings, and not to examine the relative rewards and risks of purchasing wholesale electricity will definitely see a substantial increase in their electricity costs that could wipe out their profits. Those that *do* choose to develop and implement an energy cost reduction strategy can reduce their electricity costs by up to 50% compared with the business-as-normal approach and can remain competitive and even grow their profits.

Developing an integrated approach to electricity

In order to achieve significant electricity cost savings, an integrated approach needs to be taken with an overarching energy strategic plan. This involves using less electricity (efficiency), paying less for each unit that is used (energy costs), paying less for the "fixed" network supply charges (network costs), and getting exposure to the renewable energy revolution. It also involves ongoing monitoring to lock in savings and further improvement.

These issues will be discussed throughout the book, but let's have a quick look at each of them here:

- Reducing electricity consumption includes identifying and eliminating energy use waste, investing in energy-efficient equipment and appliances, and investing in solar generation to offset use.
- Paying less for each unit of electricity involves understanding the market and getting the timing right when going out to the market to seek the lowest cost or by taking a wholesale market (spot) pool price pass-through arrangement.
- Paying less for "fixed" network supply charges involves developing an understanding of the electricity tariff components and leveraging your unique load profile to take advantage of the cost levers in the tariff.

- Getting exposure to the renewable energy revolution may involve direct investment in solar, wind or heat recovery, or negotiating a long-term offtake agreement with a renewable energy project.
- Ongoing monitoring and improvement involves detailed monitoring and reporting on half-hour energy use to ensure that load management practices are effective and to identify patterns that may lead to further load reduction and savings.

Increasing energy costs are a major strategic threat to the Australian manufacturing industry, and reducing energy costs needs to be approached accordingly. This book introduces the 9-Step Electricity Cost Reduction Framework, which incorporates each of these elements to reduce electricity costs by up to 50% in your business compared with traditional approaches, and to therefore significantly boost your profits.

But before you can develop an energy strategy, you first must understand where you are now and what needs to be improved in your business. To help with this, let's have a look at the eight most common mistakes businesses make when buying electricity.

Do any of these sound familiar to you?

CHAPTER 2

The eight most common mistakes when buying electricity

The traditional approach to buying electricity has simply been to see it as any other purchase: businesses would try to secure a good price and good terms, but there was little or no thought given to energy as a strategic issue, despite its importance to many businesses. But the traditional approach to buying electricity almost guarantees locking in much higher electricity costs than can be achieved with the more modern, innovative and sophisticated methods we are going to examine in this book. This is a huge missed opportunity for most businesses.

The eight electricity purchasing mistakes

There are eight very common mistakes made when it comes to purchasing electricity using the traditional method. These are issues I have seen firsthand time and time again:

1 automatically rolling over contracts at the end of a contract period

2 going to the market at the wrong time

3 choosing the wrong energy consultant

4 not understanding the electricity market

5 not understanding alternative electricity purchasing options (or that there are options)

6 not understanding your own electricity profile

7 not understanding your electricity bill

8 not having an energy strategy.

Let's have a look at each of these important issues.

Mistake #1: Automatically rolling over contracts at the end of a contract period

Electricity retailers can often do a great job in achieving customer loyalty. Taking customers to dinner and drinking great wine, feting them in corporate boxes to watch the cricket or football, or taking them out to shows. They build up a good relationship with their customers, as they should, and often become genuine friends. It's good customer relationship management.

Then, when the current supply agreement term comes to an end, the retailer emails a new term sheet and the customer simply agrees to the new prices, even though there might be a large increase. The relationship between the responsible manager or business owner and the retailer is so strong that there is no thought of looking at the alternatives.

The wining and dining can be great, and I myself have enjoyed sipping Grange Hermitage at some of the finest restaurants in Australia and great hospitality in corporate boxes at major sporting events. However, it may be that the difference between the new rates and other alternatives is hundreds of thousands of dollars, or even millions. That's enough to make your Grange taste cork tainted.

Unless you are ambivalent to the profit your business makes or you are an owner of the business and simply like dealing with your retail contract account manager, you should always, *always* go out to the market for pricing offers well prior to your contract term coming to an end. There are often better deals out there, or simply the threat of leaving may get you a better price with your existing supplier. Even if you have a great personal relationship with them, don't let your account manager take you – and your business – for granted.

Mistake #2: Going to the market at the wrong time

Wholesale market half-hour pool prices, or spot prices, vary throughout the year, and from year to year. Prices are often high in summer and in the depths of winter. Retail prices have a correlation with spot prices, and so when spot prices go up, retail price offers often follow. If there is a significant market event such as a supply disruption, extreme weather, drought, or a power station closure announcement, that will push spot prices up and retail prices will often follow.

It's important to understand what is going on in the market prior to putting out requests for proposals for retail supply offers. If you don't, you may well be locking in prices at the top of the market.

This is an area that an energy consultant can help you with. They should know what is happening in the market and be able to advise you about when would be a good time to seek supply proposals. However, it's better if you know what is happening in the market yourself and have your own view as to whether prices are likely to settle down or if there is a real risk of them continuing to go higher. Be aware, though, that the retailer will always have better knowledge than you. It pays to be well informed, but you can never have as much information as those on the other side of the fence.

Mistake #3: Choosing the wrong energy consultant

Energy consultants can save you hundreds of thousands – if not millions – of dollars a year. However, they could also be delivering you zero in value if you have chosen the wrong person or are not using them wisely.

A mistake that some businesses make is to engage a consultant to run an electricity tender or requests for proposals every three years and then not see them in between. You do not need to pay an energy consultant to run a tender process for you. If you employ an energy procurement manager and they engage an energy consultant to run the process then you do not need the energy procurement manager.

The energy request for proposal process often looks like this:

1. Consultant gets company to sign a letter authorising them to request load data from the retailer or network provider.
2. Consultant sends off email requesting data.

3 Consultant receives email with profile data.

4 Consultant sends email with load data off to the contacts that he/she has at half a dozen retailers.

5 Retailers send offers to consultant.

6 Consultant plays the argy bargy game with retailers to trim down margins a little.

7 Consultant puts the figures in a spreadsheet and compares offers.

8 Consultant sends spreadsheet along with formal offers to the client, including a recommendation and demonstrating how he/she pushed the retailers to achieve lower prices.

9 Consultant invoices client for $10,000.

10 Client accepts recommendation, argues terms and conditions with retailer, and signs offer.

11 Client, consultant and retailer go to best restaurant in town to celebrate deal and enjoy a bottle of Grange.

12 2.5 years later, consultant calls client to let them know that supply contract is up for renewal.

13 Repeat!

Does that seem like the consultant has added much value to the process?

If you know nothing about energy, or if energy is just one of the many categories you are managing, or if you are new to energy procurement, using an experienced energy consultant can be invaluable to help you learn. But as you learn, you should be able to understand what is happening in the market yourself and develop the industry contacts to make direct approaches for requests for proposals.

However, a good energy consultant does much more than just shuffle papers and figures for you. A good energy consultant provides you with timely information about what's happening in the market, analyses your load profile to identify opportunities for savings, gets to know your business and your operations, provides

monthly reporting on your energy cost performance, helps you to develop an overarching energy strategy, and supports you to continue to drive energy costs down year on year. A good energy consultant saves you money; they do not cost you money.

Mistake #4: Not understanding the electricity market

If electricity is a significant cost for your business then you should have a good understanding of what's happening in the electricity market. A common mistake made by many energy users is to not understand the market and then be surprised by sudden and rapid price increases.

In a similar way that you need to understand your sales drivers, revenue drivers, labour cost drivers and the cost of your raw materials, you need to understand the cost drivers of your energy bills. Understanding how the markets operate, the ever-changing dynamics of the markets, and the opportunities to exploit the markets to your own advantage is key to reducing energy costs. For example, the 2016–17 retail price increases were completely foreseeable. Wholesale prices always flow through to fixed retail prices. Anything that affects the wholesale market – such as coal-fired power station closures, increasing gas prices, droughts, new generation investment, growth in wind and solar generation – eventually shows up in the fixed retail price offers. If you don't understand what's driving the market, you would not have seen these price increases coming and would have been caught paying too much for your electricity.

It's also beneficial for the whole business community if energy users are informed and engaged. Governments have a limited understanding of how market dynamics and market imperfections are playing out specifically for consumers. They understand issues with supply dynamics and market efficiency, and know that customers are complaining about costs going up, but they are used to – and to some extent immune to – businesses complaining about the increasing cost of doing business. They are, however, very interested in informed input from knowledgeable end users. It's critical that small, medium and large end users understand the market and collectively lobby for reforms that support market-driven downward pressure on prices. There are two major energy user representative groups: Major Energy Users Incorporated (MEU) and the Energy Users Association of Australia (EUAA). It's important to be a member of one or both of these organisations to keep abreast of market reform issues and to give them strength to lobby for reform on your behalf on issues that are affecting your business.

Mistake #5: Not understanding alternative electricity purchasing options (or that there are options)

Many consumers believe that the only way to buy electricity is from a retailer at fixed prices. This is a common but fundamental mistake.

Retailers package up an electricity supply offer to a customer that guarantees the retailer a margin with minimal – if any – risk. The underlying electricity spot market price fluctuates, and that risk is measured by the volatility of prices. In order to reduce the cost "risk", customers are paying retailers a substantial risk premium without realising it. It is true that retailers can minimise their risk by taking a portfolio approach across a large range of customers, and the consumer benefits from that. But the consumer is still paying a large risk premium.

But there is another way. A customer can purchase their electricity at wholesale market prices and accept the risk themselves. They profit if price volatility is low, and pay more if price volatility is high. The customer can still purchase through a retailer but not pay the risk management premium that vanilla retail offers include. Some consumers take this approach and are either willing to accept the risk in return for the much larger potential cost savings, or they physically manage their risk by curtailing load when prices are high and enjoy low prices the rest of the time.

Some retailers understand their customer needs and structure hybrid products that allow customers to lock in fixed rates for some of their essential load and market spot prices for the remainder of their load. This enables the user to fix a price for the load that it would not be prepared to curtail in the event of high prices and allows the remaining load to be flexible to run depending on the current market price.

It is *not* necessary to buy electricity at fixed retail prices. There are other alternatives that provide greater potential savings for a manageable risk.

Mistake #6: Not understanding your own electricity profile

Believe it or not, most businesses do not really understand their own electricity use profile. That's because it's not their core business. However, understanding that profile is critical to understanding electricity cost drivers. That's why retailers need to obtain a potential customer's load profile before offering a price.

Electricity price offers are structured around peak and off-peak prices for the energy component, but there is also a network charge component around different peak and off-peak periods. Electricity costs can be decreased significantly just by shifting load to different periods or by avoiding certain periods.

Most businesses believe that they do not have the ability to curtail or shift any load. They cannot compromise their ability to meet their customers' demands, and they assume that requires the plant to run at capacity throughout their normal operating hours. I had one client who insisted that they had no ability, at all, to curtail load, and that their plant had to be kept at 100% of operational capacity. But when their data was analysed it was quite clear that they did curtail load throughout the day. They switched a large proportion of their load off at 10:00 am every day, and almost all of their load off at midday. They were obviously curtailing their load for morning tea and lunch breaks.

I had another client who stated that they absolutely had the requirement to run past 4:00 pm to meet customer demand shortfall and they could not possibly switch to a tariff that provided much cheaper prices if they they did not run past 4:00 pm. But the data revealed that they never, ever, ran past 4:30 pm, so the demand was clearly overestimated.

All operations have some flexibility. Of course the operational asset that is the plant bottleneck or constraint is usually required to run at 100% capacity, but there are almost always other loads that can be switched off if energy prices are high without impacting customer supply, quality, equipment integrity, safety or the environment.

To maximise energy cost reduction opportunities you need to understand the constraints, bottlenecks and flexibility of your operations and examine how your load profile can be flexed to minimise costs.

Mistake #7: Not understanding your electricity bill

For many businesses, the energy (flowing electrons) costs often account for only 50% of the total energy bill. Even then, energy costs are driven by when that energy is used. Network costs and other supply charges can often account for up to 50% of the total costs.

Peak and off-peak periods

When you use your energy and the maximum load when you use it are the largest drivers of energy costs, and not how clever you are at negotiating a supply deal. For example, peak prices are often up to double off-peak prices. So, the scheduling of production has a huge impact on the cost of production. It may be the case that a business may have to confine their operations to peak periods due to human resource or regulatory constraints, but operating in off-peak periods at night time and weekends is much cheaper from an energy perspective than operating in peak periods.

Network tariffs

It's not just the energy component that is cheaper. Network charges are also lower during their designated off-peak periods.

Network tariffs vary between jurisdictions, but one thing that is common is that network demand charges are determined by the maximum demand of any one half-hour period over that month. In some jurisdictions, such as South Australia, the maximum demand is set on the highest demand in any one half-hour period over a historical period greater than a year. One short half-hour spike three years ago may be driving higher costs than is necessary.

In some regions, the network tariff is set based on the amount of electricity that the business uses. In other regions there is some choice as to which tariff you use. Very large users can also have "negotiated" tariffs.

"Agreed demand" vs "actual demand"

In South Australia, there are two types of common businesses tariffs: "agreed demand" and "actual demand". With the agreed demand tariff, the maximum demand that applies to the bill every month, irrespective of the actual demand for that month, is set at the highest half-hour demand set during previous historical peak periods (November to March, midday to 9:00 pm on working days).

There is also an "additional demand" charge that is applied, which is set to the highest historical demand outside of the peak period, less the agreed demand amount. The additional demand charge rate is around half of the agreed demand rate.

With the actual demand tariff, there are three components:

- a peak demand charge
- a shoulder demand charge
- an off-peak demand charge.

The peak demand charge is based on the highest half-hour demand during peak periods (November to March, 4:00 pm to 9:00 pm, work days), the shoulder demand charge is determined by the highest half-hour demand during shoulder periods (midday to 4:00 pm, work days). Off-peak demand charges are zero.

Depending on your load profile, the agreed demand tariff or the actual demand tariff will result in lower costs. An operation that runs past 4:00 pm on working days is likely better off on an agreed demand tariff, whereas an operation that turns most of its load off prior to 4:00 pm will be better off on an actual demand tariff. However, from November to March, if a business on an actual demand tariff runs past 4:00 pm it will incur a substantial cost penalty.

"Apparent power" vs "real power"

Maximum demand is almost always measured in "apparent power" (kVA) and not "real power" (kW). Apparent power is equivalent to real power divided by the site power factor. It's not uncommon for sites to have a power factor of the order of 0.70. This means that they are paying 30% more for their demand charges than necessary to run the equipment and the load that they require. This can be rectified by installing power factor correction at a modest cost to bring the site power factor very close to 1.0. Usually power factor correction pays for itself within one to two years. (We examine power factor in more detail later in the book.)

By doing a deep dive analysis on historical half-hour demand, a business can pinpoint causes of higher demand and seek to eliminate or reduce the root causes and subsequently reduce demand charges.

Metering charges

Metering charges have also generally been thought of as fixed. Metering services are now contestable, and it's possible to achieve substantial reductions in metering charges. You can choose your own metering agent based on competitive metering rates and not just rely on the retailer to provide this service at inflated rates.

Environmental charges

Environmental charges are also a component of your bill. Your retailer purchases renewable energy certificates on your behalf and passes those costs through to your business. At mid-2017 pricing, large-scale generation certificates (LGCs) are costing an additional \$85/MWh (8.5¢/kWh) for 14.22% of the electricity that you use and small-scale technology certificates (STCs) are costing around \$34/MWh (3.4¢/kWh) for 7.01% of the electricity that your business is using. Once again, there are options. You do not have to accept the default retailer pricing.

You could also consider:

- generating your own renewable energy, large scale or small scale, to both meet your obligations and reduce your metered consumption
- purchasing your own certificates from the renewable energy certificate market
- entering into an arrangement for discounted certificates from a renewable energy scheme
- purchasing electricity through a long-term offtake agreement with a renewable energy project.

By understanding how the electricity tariff is structured, a business can make informed decisions on which is the best tariff for its profile, how to schedule operations to minimise charges, alternatives for metering agents, and options to reduce environmental charges.

Mistake #8: Not having an energy strategy

For many businesses, energy costs are in the top two or three cost inputs, up with labour. Increasing or decreasing costs will have an enormous impact on profit, yet it is not common for businesses to have a strategic energy plan to reduce long-term energy costs.

Labour costs are usually the top input cost. It is quite common for there to be a comprehensive human resource strategic plan to minimise labour costs and maximise productivity. This strategy would include industrial agreements, training and development, recruitment and retention, rewards and recognition, performance reviews and succession planning. It's difficult to put a dollar value on the benefits of this strategy as they are largely intangible. However, we all know that it's extremely important to have a human resource strategic plan or set of strategies. It is also very common to have sales and marketing strategic plans to help increase revenue, and it's quite common to have an overall business strategic plan that usually involves increasing sales and reducing costs.

It is less common to deal with energy use and procurement at a strategic level. If you wish to significantly reduce energy costs, you must have a strategic energy plan to guide you and ensure alignment and accountability throughout the organisation, and to lift the level of decision making.

This book addresses each of these common mistakes and guides you to alternatives that will help you reduce costs and increase profits.

CHAPTER 3

The 9-Step Electricity Cost Reduction Framework

A framework provides structure, logic, and a sequential process that can be followed by anyone. This 9-Step Electricity Cost Reduction Framework has been used successfully in businesses in which I have had the responsibility for managing the operations, as well as in client businesses where I have assisted in helping reduce electricity costs. It has been developed by combining an operations perspective with knowledge of how the electricity market operates, and by applying quantitative analysis and financial risk techniques.

The context for the framework is the Eastern Australian National Electricity Market (NEM), but the principles can be applied in other international electricity jurisdictions and even in Western Australia. The term "energy" is quite broad and can include process fuels such as coal, gas, liquid and biofuels, alternative waste-derived process fuels, and it can include transportation fuels such as petrol and diesel. The scope of this book is confined to electricity, but many of the principles can, and have been, applied to the other components of energy.

The framework

This chapter provides a brief overview of the framework and the structure of the book. Some readers may already have applied some or many of these steps, but there are even greater opportunities in following the framework and applying all of the steps.

The steps are:

1 Develop your strategy.
2 Analyse your operations.
3 Understand the electricity market.
4 Understand your electricity bill.
5 Reduce your network costs.
6 Reduce your energy unit costs.
7 Reduce your electricity consumption and improve efficiency.
8 Understand renewable electricity.
9 Monitor, report and continuously improve.

Let's take a quick look at each of these.

1. Develop your strategy

It may seem odd but the first step is to develop your strategy. Many businesses that spend tens or hundreds of thousands – or even millions – of dollars on energy each year have no energy cost reduction strategy. It's important to first identify your own specific energy objective, as that will drive the rest of your strategy and decisions.

2. Analyse your operations

It's critical to fully understand how electricity is used in your business. You need to be able to thoroughly and accurately answer questions such as:

- What time of day, day of week, and month of year is electricity use at its highest and lowest?
- What is the maximum demand and when does it occur?
- What drives maximum demand?

- Which equipment has what load?
- Which equipment needs to run all of the time?
- Which equipment does not run at capacity?
- What is the site power factor?

Without knowing these answers you cannot significantly reduce electricity costs. One of the common mistakes in electricity purchasing is not knowing the electricity usage profile of the operation and not understanding the operational flexibilities of the equipment.

3. Understand the electricity market

It is also critical to understand the electricity market. You must be able to answer questions such as:

- What drives electricity prices?
- How are retail prices constructed?
- What drives wholesale prices?
- What are the historical price patterns?
- When are prices high and low?
- What time of day, day of week, and month of year are prices highest and lowest?
- When is it best to go to the market for pricing offers?

To effectively play the game we need to understand the rules, participants, operation and historical price outcomes of the market, in order to understand and manage risk and seek out the opportunities.

4. Understand your electricity bill

The electricity bill will consist of energy charges based on time of use, network charges based on maximum demand and energy usage, charges for renewable energy certificates, metering charges, market charges and supply charges. Different states and even different regions within states have different tariff structures, however they have common components. Knowing exactly how these components are charged will allow an end user to monitor and reduce those component charges through load management, equipment selection, retailer selection, metering agent selection, and environmental charges.

5. Reduce your network costs

Network costs can often account for more than 50% of the bill. They are often thought of as fixed charges and so are not closely looked at. However, they are not fixed. There are a range of opportunities to reduce "fixed" network charges by reducing maximum demand charges, load profile management, or even switching to a more appropriate tariff.

6. Reduce your energy unit costs

Most customers buy electricity through vanilla retail supply contracts at fixed prices. These prices are constructed by the retailer based on expected wholesale market prices plus various risk premiums and margins. But it is possible to buy electricity at wholesale prices and not pay the risk premiums. Historical analysis shows the expected risks and rewards of purchasing electricity at wholesale prices. This step identifies those risks, the opportunities, and the risk mitigation tactics of purchasing electricity at wholesale (spot) market prices. It also provides options on hybrid fixed and spot prices.

7. Reduce your electricity consumption and improve efficiency

The cheapest electricity is the electricity that you do not use. The historical low cost of electricity drove initial investment in less-efficient technology and equipment to reduce capital costs, and has promoted wasteful practices. There are a very wide range of technologies, equipment and practices that can significantly reduce electricity consumption through waste elimination and more energy-efficient equipment. This step identifies those opportunities.

8. Understand renewable energy

Many innovative companies are investing in alternative energy sources such as solar generation and waste heat recovery. This is an important element of any energy strategy if a business is serious about reducing electricity spend. This step is often eagerly jumped to in some businesses and completely shunned by others. In this step of the framework we will deal with the business case for alternative energy.

9. Monitor, report and continuously improve

Often initiatives to increase electricity efficiency, reduce network charges and reduce unit costs gradually fade away without the business even being aware. It is critical to continuously monitor electricity consumption, demand patterns, power quality and spot price outcomes to ensure savings are both locked in and improved upon.

PART II

Demystifying electricity supply

CHAPTER 4

Understanding the National Electricity Market

As I've mentioned several times in the opening chapters, an essential part of a good energy strategy is having a thorough understanding of how the energy market works. Without an understanding of how electricity is bought and sold you have no chance of capitalising on the opportunities that arise from having a good understanding of the market. So, before we get into looking at the framework in detail, in the next few chapters we're going to delve into the inner workings of the electricity market. And a word of warning: some of these explanations and graphs get a bit technical, but none of it is too difficult and you must understand these terms and concepts to successfully navigate the energy market. If you have any problems, just re-read the relevant section again until you understand it.

The National Electricity Market

The National Electricity Market (NEM) commenced operation as a wholesale market for the supply of electricity in December 1998. The NEM supplies retailers and wholesale customers in Queensland, New South Wales, ACT, Victoria, South Australia and Tasmania, and is the world's longest interconnected power system, covering a distance of more than 5,000 kilometres with approximately $10 billion of electricity traded annually.[1]

Electricity is a unique commodity in that it is generated and consumed instantaneously and cannot be easily stored. Furthermore, it is a commodity with no differentiation between what is produced by the different suppliers. These two features make electricity ideally suited for trading via a pool mechanism, where electricity generated by suppliers is aggregated and scheduled to meet demand. The NEM is not a physical location but a set of rules and procedures managing the market participants.

The trading process

The Australian Energy Market Operator (AEMO) is responsible for managing the operation of the NEM in accordance with the National Electricity Rules to ensure that demand is met with the lowest cost supply and that electricity quality parameters such as voltage and frequency are maintained.

The wholesale trading of electricity is conducted as a spot market where generators bid specific amounts of energy at a particular price for every five-minute period of each day for each state region. AEMO then selects the bids that have the lowest price for the volume that meets the demand at that time. Those generators are then dispatched to produce the specified electricity for the successful bid volumes. The five-minute price is the highest successful bid price to meet the demand at that time. The five-minute prices are then averaged over each half-hour period to determine the regional spot price for the half-hour trading interval.

The National Electricity Rules set out a maximum spot price, the market price cap and a minimum price called the market floor price. The 2017–18 maximum and minimum prices are set at $14,200 and –$1,000 respectively, and these can be adjusted annually in line with inflation. The two key points with the market price cap and market floor price are that the cap is very, very high and the floor is negative. Yes, sometimes you can be paid to use electricity.

There is also an administered price cap (APC) if the accumulated half-hourly spot price over 336 half-hour trading intervals (one week) exceeds a cumulative price threshold, set at $212,800 in late

---

1 www.aemo.com.au.

2017. This is equivalent to an average spot price over the seven-day period of $633.33/MWh. At the time of writing, the value of the APC is set at $300/MWh until the end of the trading day on which the cumulative price for 336 trading intervals drops below $212,800.

This is a very important feature of the market as it ensures that the average price over a week will never exceed $633.33/MWh at these threshold settings. So while half-hour prices can go to the dizzying and scary heights of $14,200/MWh, the average price over a week will always be at, or more likely much below, $633.33/MWh. This price though is still approximately ten times the usual average of spot prices.

The operation of the NEM requires AEMO to determine demand levels, receive bids from suppliers, schedule generators, dispatch generators into production, set the spot price, meter electricity use and financially settle the market. The forecasting of regional demand is critical for the provision of timely market information regarding the forecast regional spot price and consequently end-user demand and risk management.

The NEM regions

There are five interconnected regions that comprise the NEM. These are Queensland, New South Wales, Victoria, South Australia and Tasmania. These regions are connected via high-voltage transmission lines called interconnectors. These interconnectors have physical limitations that result in spot price differences between regions, but nevertheless they are critical in ensuring NEM supply security and minimising prices and price volatility.

The image on the following page shows the five interconnected regions, or states, and the interconnector links. When one state does not have sufficient generation capacity to meet demand, or prices have increased due to high demand or generation restrictions, then connected regions can support that state. For example, South Australia was often supplied electricity from Victoria due to excess, low cost, generation capacity in Victoria and attractive prices in South Australia. Since the closure of the Hazelwood Power Station in Victoria, South Australia has become a net exporter of electricity to Victoria, exporting its excess wind generation.

The NEM regions

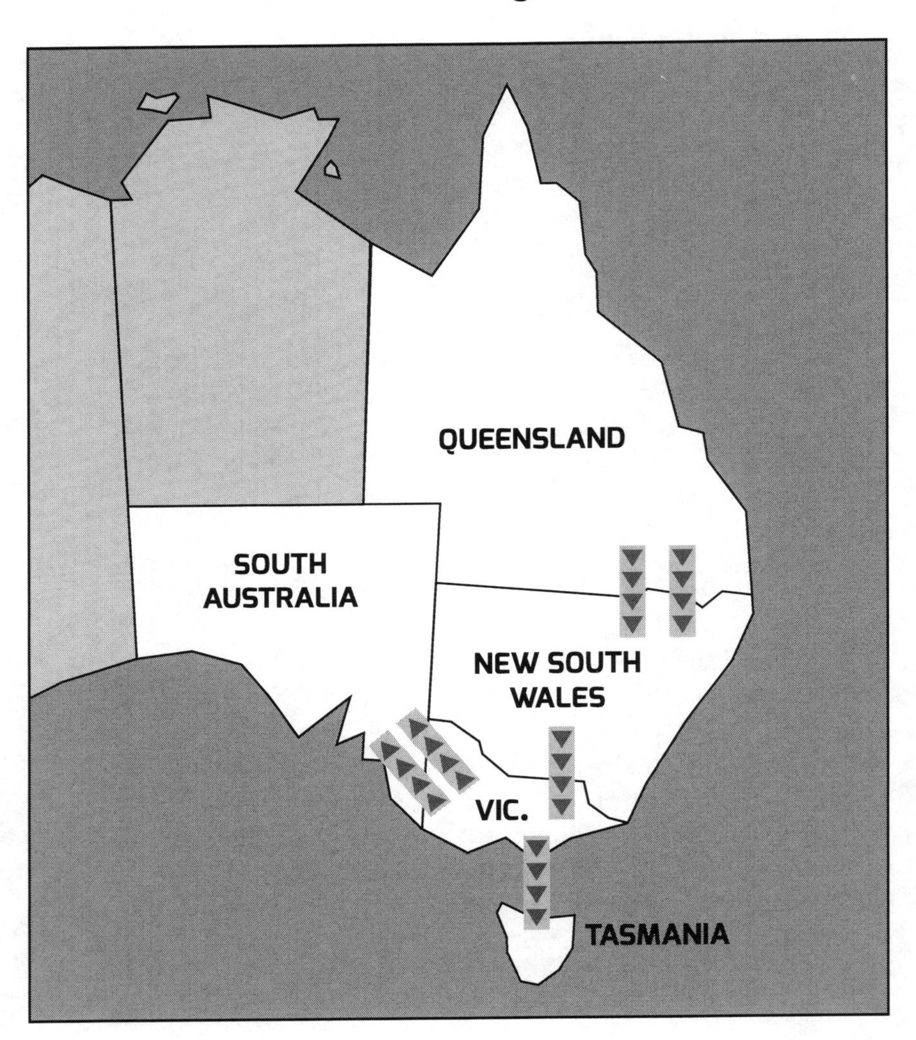

The NEM participants

The registered NEM participants are:

- **generators:** produce electricity and sell through the spot market and receive the spot price
- **market network service providers:** own and operate electricity transmission and distribution networks linked to the national grid in each region
- **market customers:** purchase electricity supplied to a connection point on the NEM transmission or distribution system for the regional spot price. Market customers are either:
  - *electricity retailers:* buy electricity at the spot price and sell it to retail customers; or
  - *end-use customers:* buy electricity at the spot price for own use.

Prior to deregulation, the four main functions of electricity supply – generation, transmission, distribution and retail supply – were performed by a single, vertically integrated state government monopoly in Victoria and South Australia. In New South Wales and Queensland, generation and transmission were contained in single state-government-owned monopolies with distribution and retail supply carried out by a number of regional monopoly franchises.

Deregulation involved the unbundling of the four functions into different businesses and providing contestability to customers; that is, the ability to choose a supplier. The classification criteria for contestability was based on energy usage, and changed to expand with time after the commencement of the NEM.

Prices in each region

New South Wales

New South Wales is the largest market. It is dominated by black-coal generation and significant low-cost hydroelectric generation, and there is significant competition. The combination of low-cost generation and high competition has kept New South Wales prices historically low. The following graph has several features that provide a great illustration of some of the factors that cause electricity prices to fluctuate. The price spike in 2007 was due to the severe drought that affected all of Australia and drove prices up in all regions due to reduced output from freshwater-cooled, coal-fired generators in Queensland and reduced generation from the Snowy Hydro system and other hydro systems. The impact of the "price on carbon" is evident from mid 2012 to 2014, as we will see in each of the regions. But the "price on carbon" has been overshadowed by the large increases in 2017. Prices jumped up in 2017 in response to the impact of the closure of the Victorian brown-coal-fired Hazelwood generator in February 2017, but have since started to drop back down as the market readjusts itself to a world without Hazelwood. The closure of the Hazelwood Power Station has had a much greater impact on Australian electricity prices than the 2012–14 "price on carbon".

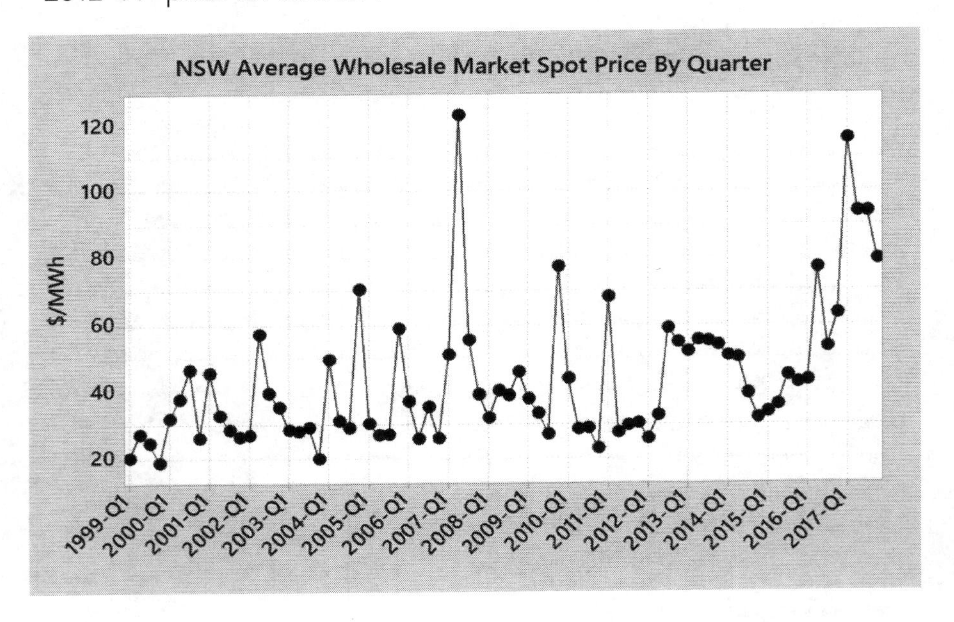

Queensland

Queensland is the next largest market. It has a high level of black-coal-fired generation. Prices have been high in the summer (March) quarter since 2013, largely due to the high level of demand associated with running compressors to pump natural gas to Curtis Island to be processed into LNG coinciding with summer peak demand periods.

Queensland had a high level of competition when the NEM commenced, with four government-owned generators plus independent generators operating independently. In 2007, this was reduced to three, and in 2011 reduced to two. The largest coal-fired generator in the state, in Gladstone, is privately owned. However, even some of its capacity is contracted to the state-owned Stanwell Corporation. The combination of low-cost generation, high summer demand, and increased market concentration in a period of high gas prices resulted in very high electricity prices in Queensland in early 2017. However, these prices have since moderated back to previous, still historically high, levels due to seasonal demand and state government intervention in their own oligopoly.

Victoria

Victoria has been dominated by low-cost brown-coal generation and a high level of competition. This has historically resulted in low wholesale market prices. The closure of the Hazelwood brown-coal-fired power station in February 2017 coupled with high gas prices saw a significant withdrawal of generation and a large increase in wholesale prices. Prices remained high, from a Victorian historical perspective, through the whole of 2017.

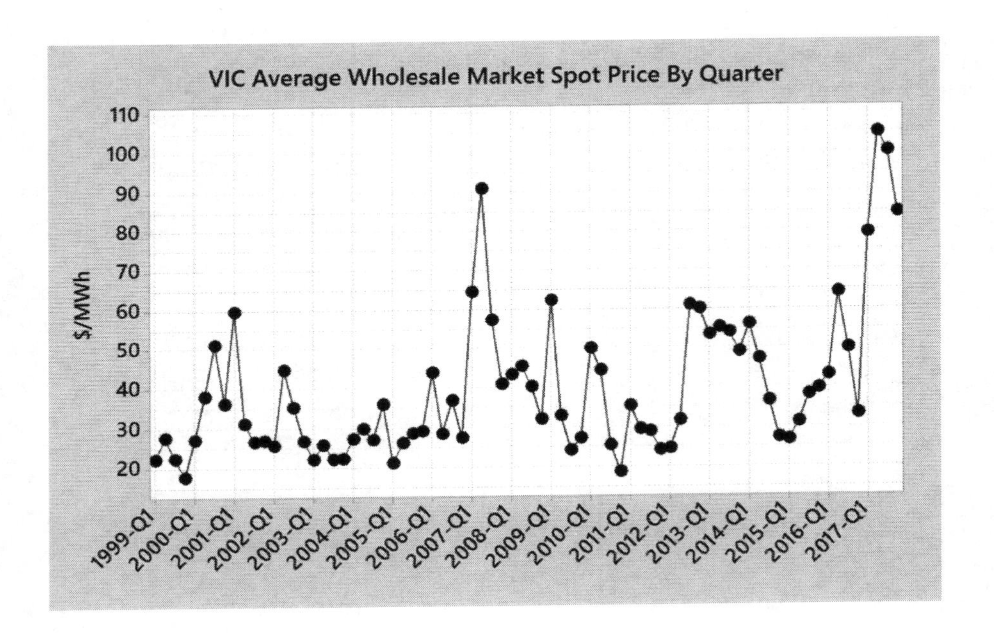

South Australia

South Australia is very interesting and unique (in addition to producing some of the best wine in the world). The energy demand is quite modest compared with the other regions and there are a small number of dominant generators. The closure of the Northern Power Station in May 2016 reduced the number of dominant generators to one, significantly reducing competition and switching the state's reliance from brown coal to natural gas and wind, supported by the two interconnectors with Victoria.. The price spiked in 2016-Q3 after the closure of the Northern Power Station was brought forward, but then dropped back down after the efficient combined-cycle gas turbine at Pelican Point was recommissioned. Prices jumped back up again after the closure of Hazelwood in Victoria, but dropped back down again with mild weather in the second half and the ramping up of the Pelican Point CCGT to full capacity.

The spot market in SA has historically been characterised by frequent and large volatility associated with fluctuations in wind generation, and the reliance on gas turbines to match wind output to daily demand fluctuations.

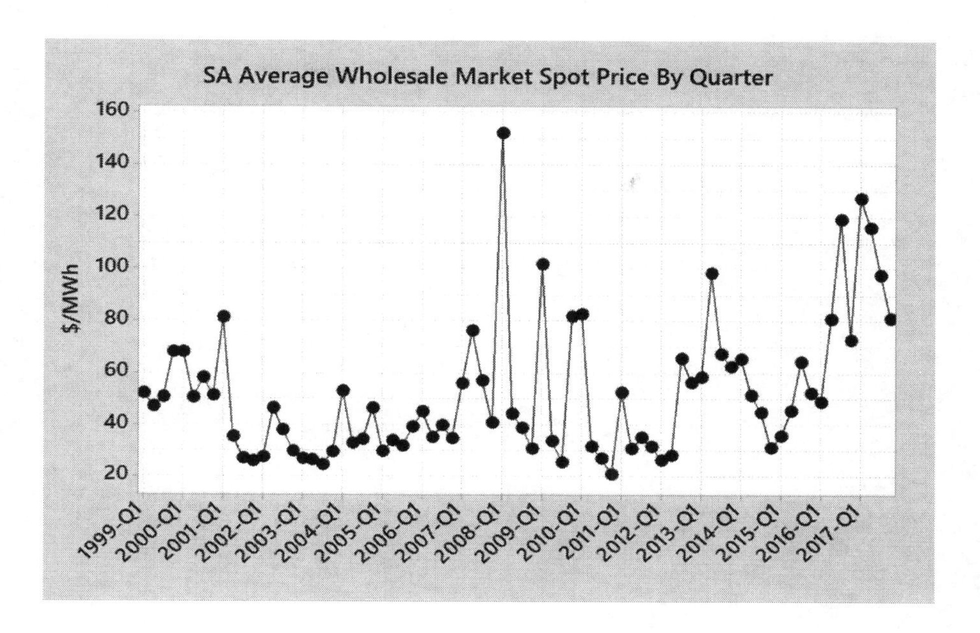

Tasmania

Tasmania has a very small demand and is largely dominated by hydroelectricity. It does have a single cable interconnector with Victoria that enables renewable and cheap electricity to be exported, and in times of low dam levels to be imported. Unfortunately in 2015 dam levels were allowed to run reasonably low with a reasonable expectation of rainfall maintaining levels. However, the underwater interconnector cable failed in 2016, which combined with low dam levels and low rainfalls to cause an electricity supply crisis in the state. Large industries were rationed by negotiation and spot prices increased sharply. Prices also jumped up in early 2017 due to the closure of the Hazelwood Power Station in Victoria, and have remained high.

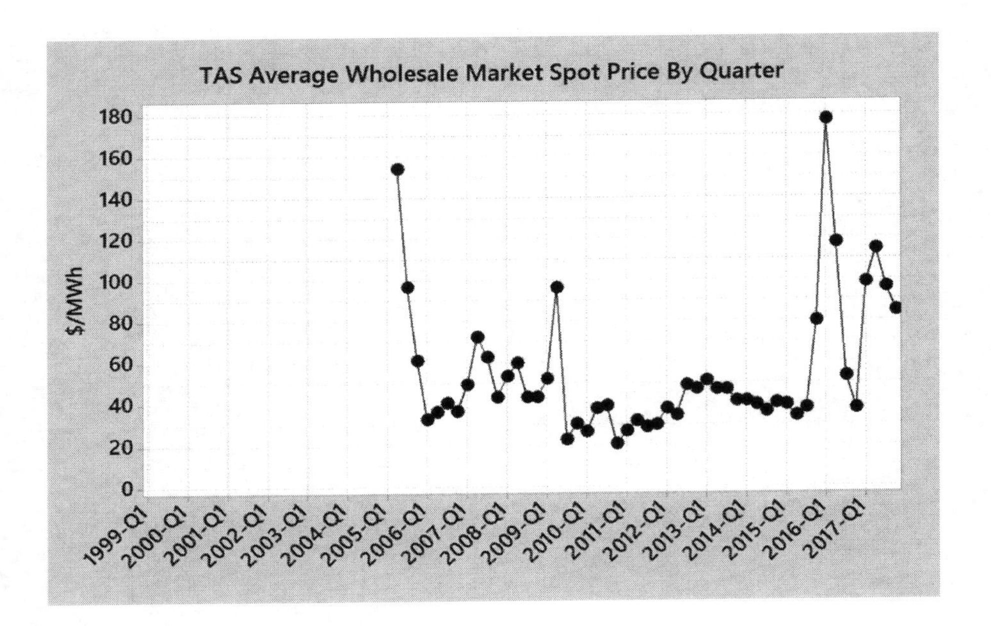

Interaction between the regions

We can see that while the NEM consists of five interconnected regions, each region has its own features that result in different risks and spot price outcomes. The interconnection between the regions is limited and constrained, sometimes catastrophically so.

What has happened from 2017 though is that there has been a shift in the average flows between regions as a result of the Hazelwood closure, and the wholesale spot pricing that was previously not highly correlated is now highly correlated. Price shocks in one region now flow into other regions.

The most stark impact has been on the flow between Victoria and South Australia. There are two interconnections between Victoria and South Australia, the larger Heywood interconnection and the smaller Murraylink. South Australia has historically been a net importer of electricity from Victoria. After the 2016 power outages as a result of severe storms, South Australia was criticised as being an electricity basket case, blindly pursuing a renewable energy ideology at the expense of energy security. Since the closure of Hazelwood, South Australia has become a net exporter of electricity to Victoria. The trends in the following graphs show the average flow between Victoria and South Australia in recent years, with a negative value being flow from South Australia to Victoria.

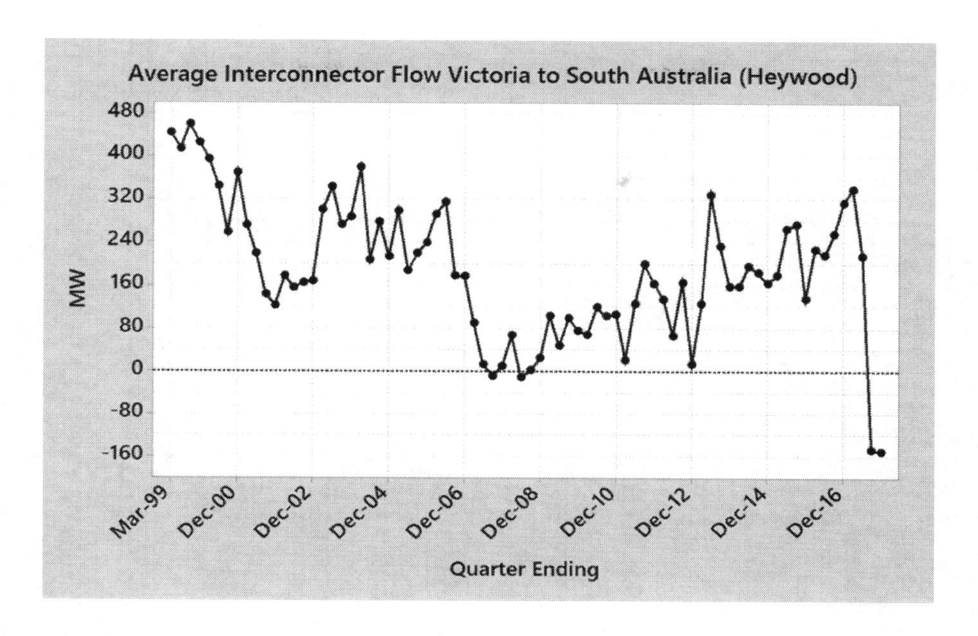

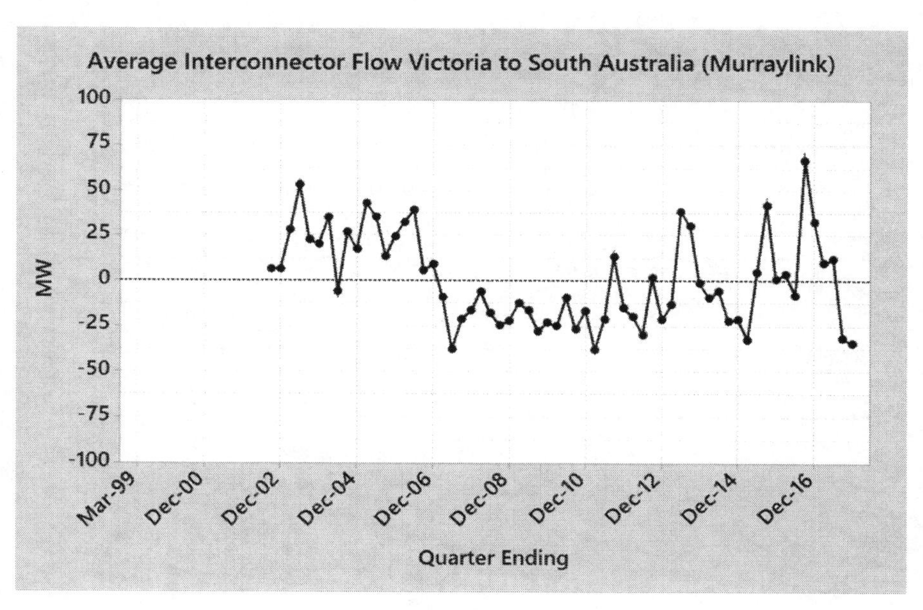

The Heywood interconnector, in particular, shows the stark change in flows from and to Victoria. The export flow from South Australia to Victoria is particularly strong when SA wind generation is high.

Average flows from Victoria to NSW have also dropped substantially compared with previous years as Victoria deals with the significant reduction in regional supply.

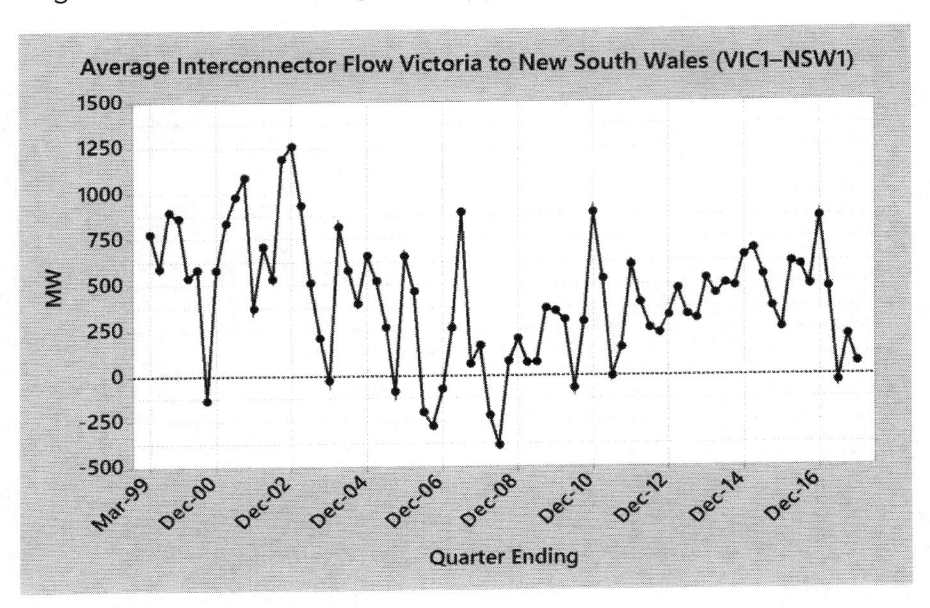

Even Tasmania has recovered from its previous low dam levels and is chipping in to help Victoria out via the Basslink.

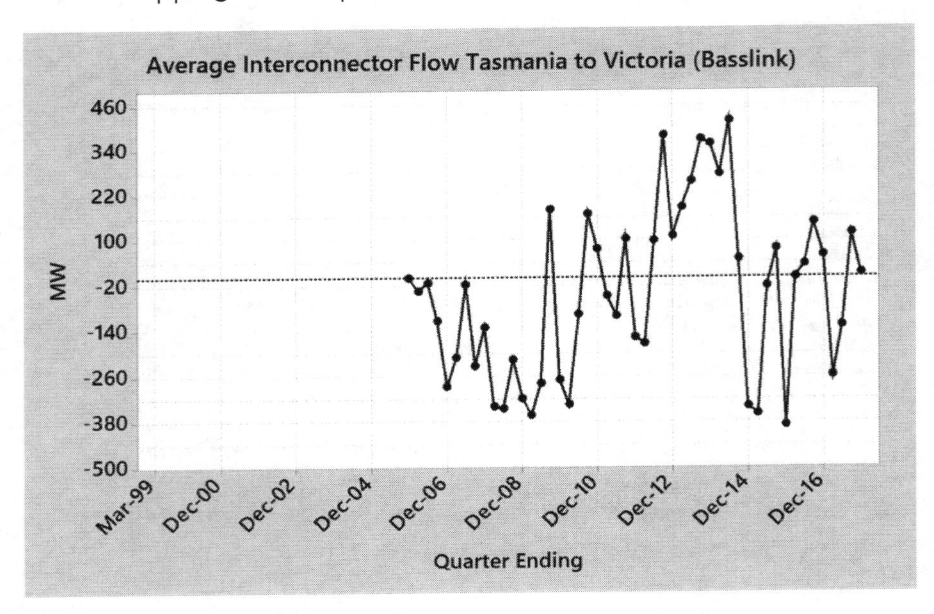

The reduction in flow of electricity from Victoria to NSW has been made up by increased exports from Queensland to NSW.

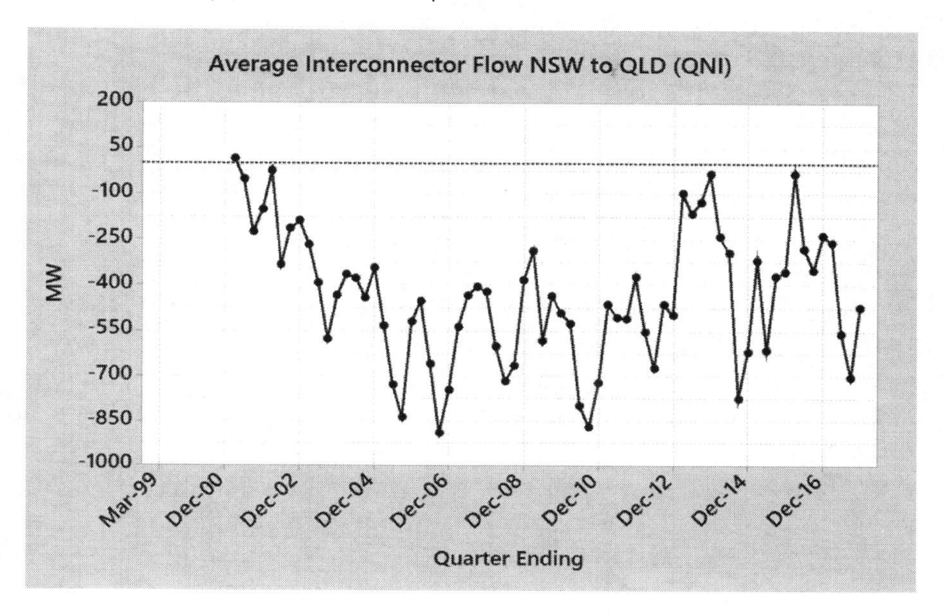

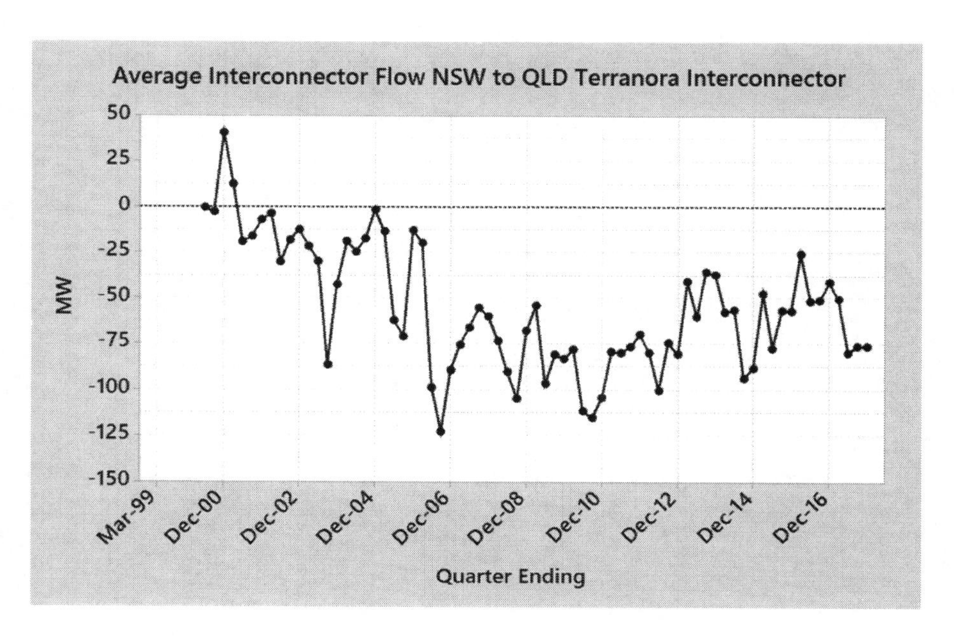

So, in 2017 we have seen the closure of the Hazelwood Power Station with an increase in wholesale prices that has spread from Victoria to each of the other states. South Australia, which used to import electricity from Victoria, is now an exporter to Victoria. Flows from Victoria to NSW have dropped substantially, so NSW is importing that amount from Queensland. This means that the high prices in Queensland due to concentration of market power, high summer demand and high gas prices are being imported into NSW.

* * *

You should now understand that the National Electricity Market is a market of connected regions, each with their own characteristics and market price drivers. You have also seen that a change in the market structure can have a large flow-on impact in each of the other regions. This understanding of the market will be very beneficial in helping you develop your own energy strategy.

CHAPTER 5

The market players

The who's who of the electricity market

To understand and successfully navigate the electricity "game" you need to understand who the players are. These players include policy makers, regulators, market operators, network providers, generators, retailers, ancillary service providers and end customers.

Market regulation and oversight

The Standing Council on Energy and Resources (SCER) is the peak body in Australia when it comes to ensuring the safe and reliable supply of electricity to consumers. It consists of the ministers responsible for energy from each of the states and territories, and the federal minister with the responsibility for energy.

Its main priorities as listed on the SCER website[2] are:

- empowering consumers
- increasing choice
- consumer protections and advocacy
- energy market transformation

2 www.coagenergycouncil.gov.au/council-priorities.

- gas supply
- gas market development
- energy and carbon policy
- energy productivity
- improving institutional performance
- review of governance arrangements
- security, sustainability and stability of the National Electricity Market.

The Australian Energy Market Commission (AEMC) is the rule maker for Australian electricity and gas markets. They also provide market development advice to the SCER.

The Australian Energy Regulator (AER) is the body that monitors and enforces compliance with the regulations determined by the AEMC. It also provides analysis of market activity and produces regular market reports, as well as investigation reports into high price events.

As discussed in the previous chapter, the Australian Energy Market Operator (AEMO) is responsible for managing the operation of the national electricity market in accordance with the National Electricity Rules to ensure that demand is met with the lowest cost supply and that electricity quality parameters such as voltage and frequency are maintained.

Transmission and distribution

Electricity is generated by a wide range of generators and in geographically diverse areas. This electricity has to be first transported from the generator to the local network where it will be used, and it then has to be transported through the local network to the end customers.

The first step where electricity is transported from the generator is called transmission. The electricity voltage has to be first stepped up to a high voltage by a transformer to minimise line losses during transmission, and then carried on high-voltage transmission lines. It is then stepped down in voltage by transformers so it can be delivered to households and factories through a transport process called distribution.

So there are transmission networks and distribution networks. These networks are either owned by the states and run as corporatised government entities or they have been privatised and sold to commercial businesses. Because this electricity transport service is a monopoly, the network owners are bound by significant regulation designed to promote efficiency and low cost of service, but also to allow them a low return based on their low risk.

Every three years the transmission and distribution networks make submissions to the AER justifying their capital investment plan and operating cost and therefore the revenue that they believe they should receive based on the regulated criteria. They then propose tariff structures and rates for each class of customer that theoretically reflect the cost of supply to each customer class.

The AER reviews each submission and makes a draft determination on the rates. The network provider then responds to the draft determination and the AER reviews and makes a final determination. This final determination can be appealed through the courts, where a judge can make a final binding determination.

What the customer ends up seeing is a tariff that is composed of both transmission and distribution charges, and is usually broken down into a fixed supply charge, a usage charge for electricity used, and a demand charge for the maximum capacity used.

The retailers

It is obvious that there are a lot of steps between generating electricity and delivering it to a customer's doorstep:

1 Generators generate electricity to meet the amount they have been scheduled by a market operator.

2 The market operator forecasts demand and runs an auction process to ensure that demand is met by the various generators.

3 Electricity is then sent down large transmission wires to local networks.

4 Electricity is then distributed through the local networks to end users.

5 Electricity consumption is measured by meters installed at the customer's supply point.

6 Customers are obligated to generate or buy a predetermined amount of their electricity from renewable generation sources.

7 The transmission and distribution networks have line losses that have to be accounted and paid for.

8 The amount of electricity that is used by end users needs to be measured and paid for.

All these different steps have to be paid for ultimately by the end users. To allow for an efficient system there is an important role provided by the retailer who bundles up the various payments into one tariff for the convenience of the customer. This eliminates the need for the end user to be a registered market participant and to pay for each of the different steps individually to the generator, transmission and distribution networks, market operator, and meter agent, and to buy renewable electricity certificates themselves. In providing this service, the retailer incurs costs of their own that they have to cover, as well as a range of risks that they are exposed to that they need to cover through premiums. They are also justified in making a margin for their services.

The distribution network company bundles up transmission and distribution charges for each different category of customer. For example, there will be residential network tariffs, small business, large business, major business and high-voltage tariffs, and many other different variants. The tariffs are designed to be cost reflective of supplying those different types of customers. The retailer passes though to the customer, at cost, the bundled network tariff along with market operator (AEMO) charges that are based on electricity consumed.

The retailer organises to either generate or purchase renewable energy certificates to meet their end customers' renewable energy obligations. They also engage a meter agent to monitor each end user's consumption so that they can be billed accurately, and they pass that metering charge through to the customer with a handy margin.

The most important function a retailer has though is to make the arrangements to supply the actual electricity. This involves studying the customer's load profile and making calculated predictions of

the customer's future electricity use, when they will use it, and how much. They then aggregate customer load profiles together into a portfolio and arrange to contract to buy the required amount of electricity from a generator or intermediary such as a bank.

There is a problem here though. Although pooling customer load profiles into a portfolio helps, retailers will never be able to accurately determine the exact amount of electricity required in each half-hour interval into the future. They may under-purchase their requirements or over-purchase. To the extent that their estimates do not match actual consumption they are exposed to the spot market prices. This constitutes a risk to them. Retailers therefore calculate their maximum risk and apply a risk premium that they pass on to the customer.

Risks for retailers

There are several risks that a retailer is exposed to:

- volume risk in the user under- or over-consuming electricity compared with their forecast
- market risk with the actual spot price that they would be exposed to with the under- and over-purchasing
- load shape risk to the extent that they have made assumptions about each client's load profile throughout the day that does not match the actual profile
- credit risk, where some customers default on payment.

Each of these risks is calculated and a risk premium is applied that is passed through to the customer.

The contracted price that the retailer pays is also a risk-premium-laden price. It is based on what the generator, which may even be the same company, believes future spot market prices will be over the contract period. Future contract prices will factor in:

- expected volatility
- changes to the market structure; for example, a power station being closed
- long-term weather forecasts
- their view of other supply and demand factors.

However, the starting point will be historical market prices.

This is an important point. Customers who choose to minimise risk and enter into a vanilla retail fixed-price contract are already exposed to the spot market. The fixed prices that the retailers offer are based on historical spot prices overlaid with a risk-laden view of future prices. They then add on the other risk premiums, their own costs, plus a margin.

Intermediaries

Another important participant in the cost-efficient supply of electricity is intermediaries. Intermediaries are often banks that purchase wholesale blocks of electricity from generators at what they view as a low or fair price, and they then sell them to retailers at prices that reflect their view of the future market. They help to provide liquidity and competitiveness to the wholesale market. To some extent they are setting the wholesale market price for electricity. Historical spot prices strongly influence wholesale contract prices that then flow into retail prices.

Some generators may not have a retail arm and so will sell their output to an intermediary. Even generators with a retail arm will also sell to intermediaries. Retailers without a generation arm can then buy electricity from an intermediary, or even some that have a generation arm may buy from intermediaries if the price is more competitive.

The Australian Securities Exchange and Austraclear

Usually the contracts are sold in quarterly blocks and they are sold as swaps or contracts for difference. This means that payment is made with reference to the spot market settled price for the quarter. This transaction is sometimes called a hedge.

The trading for swaps occurring in the marketplace also creates standardised products for which futures contracts can be constructed. A futures contract is a contract for delivery of a product at some specified time in the future. In this case it's a financial contract without delivery of a physical product. It is an exchange-traded contract for a block of specified electricity, over a specified period (month or quarter), in a specified region. The Australian Securities Exchange (ASX) operates an energy market

exchange through which futures, options and other derivatives can be traded.

The price paid for the futures contract will be the current trading price for that standardised block of electricity for some period in the future for a specific market region. The contract is settled at the end of the period. The purchaser gets paid or pays the difference between the purchase price of the contract and the settlement price for the underlying spot contract at the end of the period.

There are also other derivatives that are used, such as options and caps. With a call option, the purchaser pays a premium upfront and has the right – but not the obligation – to exercise the settlement of the underlying futures contract at the end of the period. If the average market spot price settles above the underlying futures contract price then the purchaser would likely exercise the option to lock in the futures price. If the average market spot price settles below the futures price then the purchaser would not exercise the option and would take the lower actual market price.

A cap is a product that can be purchased that sets the maximum price for any half-hour period over the quarterly block for a given load. For example, a purchaser can buy a $300 cap, which is a contract that is financially settled based on the actual half-hour prices for the quarter. It means that the purchaser will be paid the difference between the actual average price for the contract period and the price that would have occurred if the maximum price in any half-hour period was $300/MWh. This payment will offset the amount they actually paid for the spot electricity purchased inclusive of the price spikes above $300.

There is a large amount of cashflow going backwards and forwards between different market participants outside of the ASX, and in order to reduce credit risk and ensure maximum efficiency in cashflow payments, all NEM participants are required to register with Austraclear and settle their net cashflows on a prescribed weekly or monthly basis.

The Clean Energy Regulator

The Australian Government has mandated that by 2020, 33,000 GWh of electricity will be generated from renewable sources such as hydro, tidal, wind and large-scale solar. The government sets targets on how much electricity must come from renewable sources each year in order to ramp up to this 2020 target over time. Electricity users, mainly through their retailers, are required to purchase renewable energy certificates for the mandated target proportion of the electricity that they have consumed and then surrender those certificates. The Clean Energy Regulator operates an exchange where renewable energy certificates are registered by application from renewable generators who prove the generation of the renewable electricity. The certificates can then be bought and sold on the online exchange. Wholesale electricity purchasers such as retailers can purchase their required certificates through this exchange.

* * *

In order to get electricity to the end user it can be seen that there are a large number of functions to ensure that it is done safely, reliability and financially efficiently. In summary, this includes the SCER, the MCE, the AER, AEMO, generators, transmission network providers, distribution network providers, metering agents, retailers, intermediaries, Austraclear, the ASX and the Clean Energy Regulator.

CHAPTER 6

The end users

The National Electricity Objective, as stated in the National Electricity Law, is:

> *to promote efficient investment in, and efficient operation and use of, electricity services for the long term interests of consumers of electricity with respect to – price, quality, safety, reliability, and security of supply of electricity; and the reliability, safety and security of the national electricity system.*

" ... for the long term interests of consumers". Those pesky little consumers are an annoyance to the industry, and that particular phrase in the objective is – in the words of one of the architects of the market – "unfortunate".

The whole market exists to serve consumers. However, quite often it seems as if the whole market exists to serve the industry. If you are lucky enough to attend any government market reform workshop you will see that the room is dominated by players on the supply side: retailers, generators and network providers. Many times I have heard the chirping of crickets after I have contributed to the discussion by interjecting with, "Surely the ultimate outcome needs to be reduced costs for consumers".

The different types of energy users

End users consist of households, small to large businesses, government entities and industrials. Each of these end-user categories has different characteristics and load profiles. Together, in aggregate, they make up the demand profile for a given market region.

Households

With households, demand starts to increase at 6:00 am as people wake up and turn on their lights, heating or cooling, and appliances. After 9:00 am demand starts to fall as the kids go off to school and mum and dad to work. Demand picks up again at 4:00 pm when the kids start to get home, the heating or cooling are turned back on, the large TVs are turned on, and the dinner is put in the oven or on the stove (or in the microwave). Demand drops away again after 10:00 pm when everyone starts to go to bed. There is still residual demand overnight as many appliances continue to draw power in standby mode and some households keep heating or cooling on.

Businesses

Small to large businesses have a different profile. Demand starts to pick up after 8:00 am when offices or shops are opened up, heating or cooling is turned on, lighting is turned on and computers are fired up. This load remains fairly flat throughout the day, unless the weather is very hot or very cold, and then starts to drop again after 5:00 pm when the business hours come to an end and the lucky ones start to go home.

Large office buildings still have a significant after-hours load, with some lighting, heating and cooling continuing to operate overnight. There have been major advances in smart building energy management that have significantly reduced this overnight load.

Industrials

The industrials can have quite different load profiles. Some large industrials operate 24 hours a day, seven days a week, and have a fairly flat load profile. Some operate seven days a week but only on

day shift, and others operate Monday to Friday either around the clock or on day shift only.

Interactions with the market

These end users interact with the energy market in different ways.

Households and small businesses typically have a manually read meter where their consumption is measured on a quarterly basis. They are charged based on their consumption, with different prices for different levels of consumption. These consumers usually barely understand the market and quite often don't even spend the time to find the lowest cost retailer. The retailers usually set their energy tariff rate on an annual basis and simply inform the customer of annual increases. The smarter households use an energy broker to set them up with the lowest cost retailer, or they can search online for themselves. The really clever consumers negotiate with several retailers to get the best deal that suits them.

Larger businesses will often use an energy broker or seek competitive offers to supply themselves. Even if electricity is not one of their top 10 costs, they are a bit more energy conscious and often have active programs to reduce their energy consumption through energy-efficiency programs and installation of solar generation.

The industrials are often much more sensitive to energy costs. Often energy costs constitute 20% to 30% of their total costs, and they are usually in the top five costs. As such, you would usually expect them to be more innovative and relentless at reducing energy costs through smart management, energy-efficiency measures and investment in technology to reduce consumption. But quite often they are not, as they use more traditional methods of energy procurement such as tenders every three years, and the purchasing analysis and decisions are done solely by the centralised procurement department. They often lead the charge in terms of engaging formally with government to take action on rising energy costs as they have the funds to support energy-user advocacy on their behalf along with other large companies, or the depth of knowledge to engage with government directly themselves.

In general though, few household consumers and businesses have a thorough understanding of how the electricity market works and how to significantly reduce these costs. Some do, and they are reaping the rewards of increased profits and reduced costs already. The 9-Step Electricity Cost Reduction Framework in the next part will show how you can increase profits in your own business.

Understanding electricity demand profiles

Electricity demand profiles are very important in designing the network system.

There are many types of businesses, each with their own characteristic electricity usage patterns. Some businesses run 24 × 7, while others run only during business hours Monday to Friday. Many others are somewhere in between.

Later in the book when we look at implementing the electricity cost reduction framework we will see how different load profiles will result in significantly different price outcomes, and that strategy will be different depending on those profiles. The six load profiles that we will use as a basis for analysis are:

1. 1 MW flat load 24 × 7
2. 1 MW load 8 hours per day Monday to Friday
3. 1 MW load 16 hours per day Monday to Friday
4. 1 MW load 24 hours per day Monday to Friday
5. 1 MW load 8 hours per day 7 days per week
6. 1 MW load – off-peak operation (nights and weekends).

Let's have a look at each of these.

1 MW flat load 24 × 7

The flat load is typical of a large manufacturing business that operates 24 × 7. They have a price advantage in that a significant portion of the electricity usage is during off-peak periods and so their overall unit price will be lower. These businesses tend to be the larger industrial companies such as cement, steel, aluminium, metal refining and glass operations.

1 MW load 8 hours per day Monday to Friday

Businesses that operate during normal business hours have the disadvantage that they are fully exposed to peak electricity prices and don't have the advantage of off-peak electricity prices. Out of all of the load profiles, this profile will have the largest unit cost of electricity. The businesses tend to be small to medium users.

1 MW load 16 hours per day Monday to Friday

These businesses do have some exposure to off-peak prices late in the afternoon shift but are still largely facing peak-period pricing. These businesses tend to be medium size manufacturing businesses.

1 MW load 24 hours per day Monday to Friday

These businesses are seeing lower electricity prices during the night shift of their operations, bringing the overall average unit price down somewhat.

1 MW load 8 hours per day 7 days per week

These businesses have exposure to the lower off-peak prices over the weekend that reduce their overall electricity unit prices.

1 MW load – off-peak operation (nights and weekends)

These businesses maximise the opportunity of using electricity when prices are at their lower off-peak level. They tend to be businesses such as irrigators and water utilities.

* * *

Later in this book when we look at applying the framework you will see the relative impact on savings with each of these different business load profiles. It's very important for you to understand the load profile of your business and how this affects your energy costs.

CHAPTER 7

Bringing it all together

The National Electricity Market wholesale commodity exchange

The "game board" that brings the electricity suppliers and users together, through their retailer, is the National Electricity Market wholesale commodity exchange. This exchange is where prices are determined through the intersection of supply and demand. It's important to understand how prices are determined in order to leverage that information to extract price reductions. In this chapter you will discover how the electricity price auction works, why different generators bid at the auction at different prices, and why prices can be negative and very high.

The bidding process

The National Electricity Market is composed of five interconnected sub-markets that have their own regional auction process to determine the half-hour spot price. In each region the spot price is established by matching the actual demand to a bid stack of generation offers made by generators operating both within the region and outside the region.

Generators make supply offers for blocks of energy at different prices. In theory, with an efficient competitive market the generators will bid in energy at or slightly above their marginal cost of production, also known as the "short run marginal cost" (SRMC). If their offer is accepted and the generator is dispatched then they will make a profit equivalent to the difference between their SRMC and the spot price. In this way the cost of supply of electricity to meet the demand is the lowest cost of supply.

A bid stack example

In the simple example in the following chart a bid stack is constructed from offers made by generators and the settlement price is determined by the lowest cost generation required to meet the last unit of demand. In the low-demand scenario the settlement price is –$1,000/MWh, in the medium-demand scenario the settlement price is $50/MWh, and in the high-demand scenario the settlement price is $8,000/MWh.

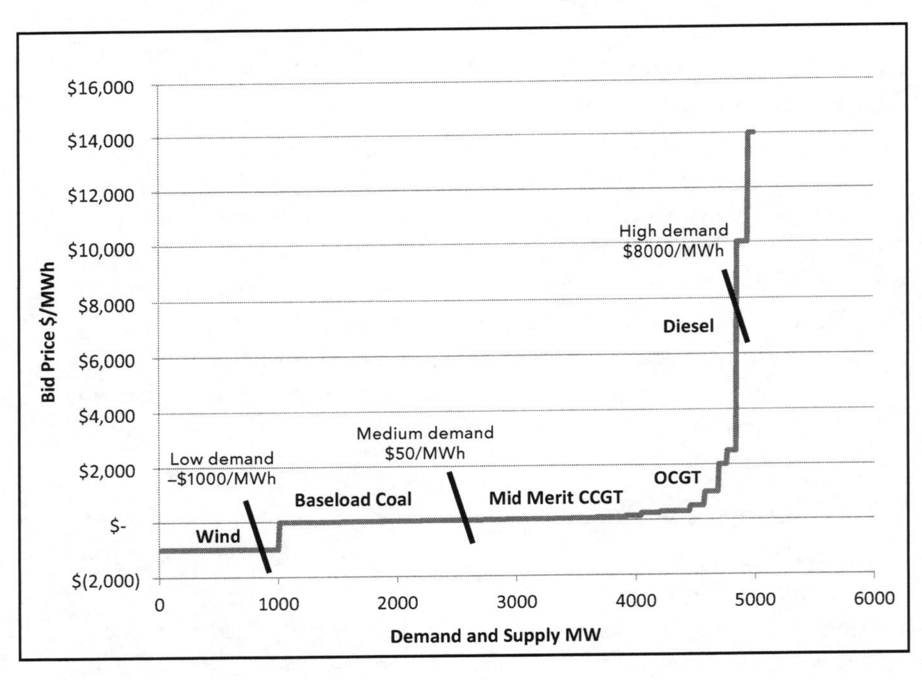

The cost effects of generation output

However, in reality it's far more complex than this. The major generation technologies are limited in their ability to "turn down" their generation output. Most coal- and gas-fired boiler steam generators have a maximum turndown of 30% to 40% of their maximum load, and combined-cycle gas turbines (CCGT) that rely on the waste steam from the first-stage gas turbines can usually turn down to 40%. Open-cycle gas turbines (OCGT) can ramp up and down between their maximum and zero very quickly.

In addition, for boiler longevity it is far better to run a boiler at a steady rate rather than thermally cycling it up and down. It also takes four to five hours to restart a warm boiler and ramp up to full rate. It's therefore quite common for coal-fired generators to bid in a negative or zero price to ensure that they are dispatched and so will not be required to ramp up or down, particularly beyond the maximum turndown.

The cost effects of generator supply contracts

Another reason for generators to bid in costs below their SRMC is because of their contract position. Generators have supply contracts with retailers and intermediaries to meet. If they are not dispatched and their output is below the supply contract requirement then they will have to pay the counterparts the difference between the spot market price and the contract price. This may be beneficial but it does leave them exposed if the spot market spikes and they have a supply shortfall. Most often the generator will want to ensure that their price bid strategy ensures they meet their contracted position.

Many generators will have part of their generation portfolio uncontracted. If they can be dispatched at a price above their SRMC then they will make a margin. They will develop a bid strategy that maximises the probability of dispatch at the highest possible price. For example, a generator with four boilers may have three fully contracted and bid three boilers' worth of capacity in at their SRMC. They then may look to get much higher prices to run their fourth boiler and make a large windfall margin if dispatched. They may also run all four boilers, each turned down 20%, and bid the unused capacity in at very high prices up to the market cap price of $14,200. As these boilers are already running it is easier for them to achieve the higher output within a shorter period of time.

Market fluctuations

Open-cycle gas turbines are quick to start up and ramp up and down. They can achieve full load from start in 10 to 15 minutes. These very important generators ensure system security by quickly ramping up and down to adjust the market supply for wind fluctuations, other generators tripping, variable flows across the interconnectors, and predictable rapid increases in system demand in the morning and late afternoon peak periods. OCGTs can bid in very high prices and make very large margins, depending on their own price of gas. They are usually setting the price when price spikes occur.

And then there is the wind and the sun. Wind and solar generation is largely automatically dispatched, in contrast to generators that need to have their bid accepted to be dispatched. Wind generation and large-scale solar are automatically bid in at a price of –$1,000/MWh. Small-scale solar simply results in negative demand; that is, it reduces the overall market demand. If renewable generation equals or exceeds the demand then the spot price will be –$1,000/MWh and end users who are exposed to market prices are paid to consume electricity during these periods. Large base-load generation such as coal-fired generation will have to stop their output during these periods but spend money to keep their boilers warm. To ensure that they are the first non-renewable to start up or the last to shut down they bid prices into the market below their short run marginal cost, SRMC, and sometimes even at negative prices. The SRMC is the incremental additional cost of producing an additional MWh of electricity, in this case, largely the cost of coal per MWh produced.

The market price for a half-hour interval can therefore fluctuate between the floor price of –$1,000/MWh and the market cap price of $14,200/MWh. This means that market-exposed consumers can be paid to use electricity if they are consuming load when prices are negative but they can also be exposed to very high prices when the market price spikes in the direction of the market cap price.

The market cap price

The market cap price is the thing that most scares end users when looking at buying electricity at wholesale prices. They either think that being exposed to that price will instantaneously wipe out

annual profits or they think that they would not be able to run equipment during these price spikes. The mistake commonly made is that they look at individual half-hour prices and not at the average monthly price in comparison to their best retail price. The retail prices have already factored in the forecast high price spikes. Consumers are already paying for them.

* * *

In chapter 10 – which covers Step 3 of the framework, Understand the electricity market – we will have a closer examination of historical market price outcomes, what drives those outcomes, and what the patterns are. This will help the reader put the high price spikes into a longer term context and understand that they are not as scary as they seem.

In chapter 13 we deal with Step 6 of the framework – Reduce your energy unit costs – by taking advantage of half-hour spot market pricing rather than one- to three-year fixed retail pricing.

PART III

Applying the framework

CHAPTER 8

Step 1: Develop your strategy

Almost all successful businesses have a strategic plan. Without a strategic plan the business cannot be clear on its objectives, priorities, the supporting actions and the accountabilities required to fulfil those objectives. Without a strategic plan businesses just accept that things "happen to them" and they are not "in control of their own destiny".

Similarly with energy. If energy costs are a significant cost item, unless you have an energy strategic plan, energy cost increases will just "happen to you". By developing a strategic plan you can set ambitious energy cost reduction objectives and then achieve them. You can take control of your energy spend and be in control of your own destiny.

So, let's get into it …

All relevant stakeholders must understand what's going on

The importance of educating the relevant stakeholders in how the energy market works cannot be underestimated. When the strategy is formulated by a team of people, each of whom brings their own skills, experience and niche subject matter expertise to the table, if they all have knowledge about how the market works then they can apply their skill set, experience and expertise in making connections. Opportunities arise by making those connections.

My own experience was like that. We didn't initially have a framework and didn't understand the market very well. We understood how the spot market worked, but as it was at the start of the market we didn't have much historical data. As I learned more and more about the whole electricity supply chain and developed experience with the spot market, other cost reduction opportunities that were previously not understood by anyone popped out. As I shared my learnings with colleagues they made other connections from their own skills and experience and highlighted further cost reductions.

For example, a team that I worked with saw the impact that very sporadic increases in demand caused to their annual demand charges. One of the production engineers said he had no idea that these very rare spikes were causing demand charges to increase by $20,000 a year because he was not aware of how electricity network charges were calculated. This issue was easily fixed by a bit of coding in their process control system.

Another team quickly made the connection that putting all of their electric forklifts on charge at the same time at the end of the day was causing their demand charges to be $10,000 a year higher than they would otherwise be if they waited until the off-peak period to charge them. A simple timer easily fixed that problem.

A good idea is to hold a training session ahead of an energy strategy formulation session so that the participants understand, at least in general, how the market works. This allows the strategy session to flush out the opportunities rather than being used as an education session.

Who should be on the team?

That brings us to who should participate in the strategy sessions.

The business owner or business unit manager should attend, as ultimately they need to buy into and own the strategy. The most difficult part of any strategy is the implementation, and so the boss of the business needs to own the strategy and ensure the buy-in of the other participants. The business owner or business unit manager needs to attend the strategy formulation session.

Many medium to larger businesses will have a procurement person who is responsible for energy. Procurement people are

usually very commercially savvy and often focus on pleasing their customer, the business unit. They usually however do not understand the intricacies of the operations: which pieces of equipment are bottlenecks, which are under-utilised, which are inefficient, and the flexibilities around when equipment needs to be operated. They are good at running a commercial process to achieve the lowest cost option. However, they usually are asking for and getting the vanilla options. The procurement person will ultimately manage the supply contract and needs to attend the strategy session.

The operations manager or engineer would like to have the lowest cost electricity, but they don't really want to be bothered with having to mess around with analysing the different options as they have other more pressing issues to worry about. They often have no knowledge of or interest in the tariff – they just want the procurement person to get the lowest price. However, they have a wealth of knowledge and experience that needs to be tapped into to understand the opportunities available to the business. It is therefore crucial that they attend the strategy formulation session.

Many medium to larger businesses have one or more electricians or electrical engineers who have a strong understanding of the electrical infrastructure of the site and the process control system. They also sometimes have no knowledge of the tariff, however they usually become very interested when they see and understand it. On many occasions I have seen the "light come on" when the tariff is explained to them. It is critical that a key electrical person attend the session.

A financial person is also an important contributor to the team. They load the numbers for production, energy consumption and cost into complex spreadsheets that only they can follow, and produce data and trend graphs for management reports. They understand the numbers, and can both contribute and build the knowledge gained into their complex and unwieldy spreadsheets or into SAP.

The accounts payable person is often the forgotten person in the organisation who just goes about their thing – processing and paying invoices. They are however a crucial gate keeper. They see the supplier invoice, check it, seek approval for payment from the business unit manager or operations manager – which is usually just a rubber stamp – and then pay it. They physically see the invoice and are the ones who can best pick up invoice errors and recognise unusual variances. Their buy-in on the strategy and ability to make connections is very important, and they should be invited to attend the strategy session.

Maintenance people plan and schedule equipment downtime and they understand the equipment thoroughly. Having a maintenance supervisor or planner attend the session and a mechanical engineer is hugely valuable to the strategy process. They can make connections, and offer to shift maintenance timing or look to replace high-electricity-consumption equipment with more efficient equipment.

Finally, the unsung productivity multiplier of the organisation – the administration person – would be an invaluable person to attend the strategy session. Not only can you utilise them to document the strategy session and track and manage actions, but they will often be extremely enthusiastic due to their being invited to participate.

A cross-functional team of the business unit manager/owner, operations manager or engineer, procurement professional, electrician/electrical engineer, accounts payable, maintenance supervisor, mechanical engineer and an administration person is a great team to include in the strategy formulation session.

Developing your strategy

There are many different ways to develop a strategy. The approach that I have found useful when working with clients is a straightforward approach using simple but very powerful tools. Don't be fooled by the simplicity of the tools – the greater the simplicity, the greater the power.

A great way to develop your strategy is:

1. Form a TEAM.
2. Clarify the OBJECTIVE.
3. Know the NOW.
4. Determine the FUTURE.
5. Flesh out the HOW.
6. Formulate ACTION.

Let's see how this all works.

1. Form a TEAM

Identifying electricity cost reduction opportunities will not just rest with an "all-knowing" senior manager who decrees what savings are possible and how they should be achieved. It requires cross-functional knowledge to tie ideas and pieces of a jigsaw puzzle together. The team should consist of a leader with organisational clout and representatives from production, maintenance, technical, procurement, finance, accounts payable, administration and any other functional area, as discussed in the first part of this chapter.

2. Clarify the OBJECTIVE

In formulating an energy strategy it's important to first determine the OBJECTIVE of the strategy. This may sound obvious but it's a step people often miss. The objectives of different organisations will vary with their values, appetite for "risk", their level of innovation, their financial exposure to electricity prices, the ownership structure, and the attitude of the owner or CEO toward forging their own future rather than waiting for things to happen to them.

Simply "wanting to reduce energy costs" is not a very helpful objective and will not set you on a path to success. More useful objectives could include:

1 Reduce overall energy consumption by 30% with smart investment in energy-efficient technologies by June 2018.

2 Reduce overall energy consumption by 10% by being more efficient without capital investment by June 2018.

3 Reduce energy costs by 50% by June 2018.

4 Secure fixed-price energy supplies for the next three years at the lowest possible price by June 2018.

Objective 1 shows a bias toward energy-efficiency improvement with a positive view on using capital investment to achieve cost savings, whereas objective 2 also has a bias toward energy efficiency but requires the savings to be found without capital investment.

Objective 3 shows that all options to reduce spend are on the table, whereas objective 4 reveals a "safe" "business-as-usual strategy".

The objective of this book is to assist businesses to reduce energy spend by up to 50%, so the framework that we will use will keep all of the options on the table. You can, however, obviously choose the objective that best suits your business.

The objective may be mandated from "above" or it may be necessary to develop one from a blank sheet. The objective should be stated in the SMART (Specific, Measurable, Achievable, Relevant, Time-bound) format. Using a SMART format helps to clarify the objective for everyone.

If you are developing an objective from a blank sheet then a great tool to use is to simply brainstorm objectives and then use your own process to distill that down to one objective.

3. Know the NOW

After you've set your objective, the next thing that needs to be established is what is happening NOW. Much of this work should be developed prior to the workshop and presented at the workshop as Step 2 of the framework: Analyse your operations. The now should include the following metrics:

- data on electricity consumption in terms of total, peak and off-peak, and total annual consumption
- data on production so that you can quantify electricity use in terms of unit production
- trend data on your load profile by time of day, day of week, and month of year
- electricity bill breakdowns – quantify each component on an annual basis
- cost per kWh (or MWh) by peak, off-peak and average
- annual cost
- maximum demand
- power factor.

The NOW might look like:

	NOW
Efficiency	100 kWh/unit
Electricity unit cost	$90/MWh
Environmental costs	$144,834
Network costs	$937,166
Metering costs	$18,000
Maximum demand	1.2 MVA
Power factor	0.833
Annual consumption	10,000 MWh
Annual electricity cost	$2,000,000

4. Determine the FUTURE

This next step involves a brainstorming session to determine your desired future state. My experience has shown that an initial timeframe of two years is realistic for this future state. The first year generally involves the measurement, analysis and implementation stages. While there can be immediate cost-saving benefits, the full annualised benefits are usually seen in the second year.

The FUTURE will be derived from your OBJECTIVE. Are we going to seek to achieve the savings through energy efficiency alone, exposure to wholesale prices, capital investment, investment in renewable generation, or all of them combined?

Our brainstorming question would be "Where do we need to be to achieve our objective?" If we are embarking on an aggressive strategy, the result could look like:

- 5% improvement in energy efficiency per unit of product
- 22% reduction in electricity unit costs
- 70% reduction in environmental costs (excluding cost of capital investment)
- 30% reduction in network costs (including power factor to 0.98)
- 67% reduction in metering charges.

Our FUTURE would look like this:

	NOW	FUTURE
Efficiency	100 kWh/unit	95 kWh/t
Electricity unit cost	$90/MWh	$70/MWh
Environmental costs	$144,834	$36,209
Network costs	$937,166	$656,016
Metering costs	$18,000	$6,000
Maximum demand	1.2 MVA	1.0 MVA
Power factor	0.833	0.98
Annual consumption	10,000 MWh	9,000 MWh
Annual electricity cost	$2,000,000	$1,328,225

This FUTURE has a $672,000 saving compared with the original $2,000,000 spend to achieve the same production output. This represents a 34% reduction in electricity costs and $672,000 straight to the bottom line.

5. Flesh out the HOW

This HOW step identifies the strategies to move from the NOW to the FUTURE. You are going to achieve these top metrics using the results of the following five steps in the framework, so you'll only be able to fully complete Step 1 after you've read all of the framework:

- reduce network costs (Step 5)
- reduce electricity unit costs (Step 6)
- reduce electricity consumption (Step 7)
- renewable electricity generation (Step 8)
- monitoring, analysis and reporting (Step 9).

I prefer to use a mind map to do this brainstorming, and have common branches around the pre-identified steps. On the following page is an example of a typical mind map.

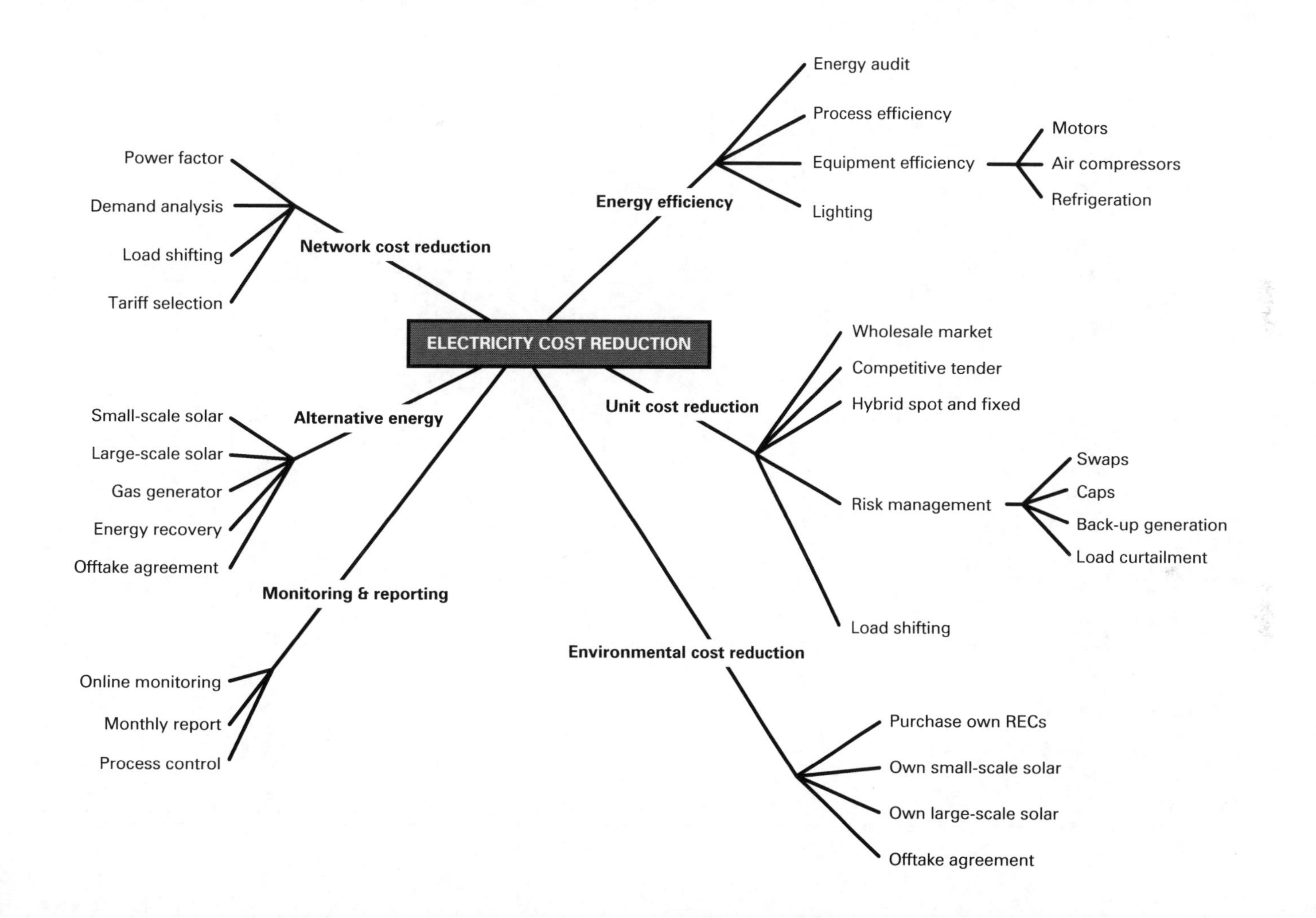
ELECTRICITY COST REDUCTION
Network cost reduction
Power factor
Demand analysis
Load shifting
Tariff selection
Energy efficiency
Energy audit
Process efficiency
Equipment efficiency
Lighting
Motors
Air compressors
Refrigeration
Alternative energy
Small-scale solar
Large-scale solar
Gas generator
Energy recovery
Offtake agreement
Unit cost reduction
Wholesale market
Competitive tender
Hybrid spot and fixed
Risk management
Load shifting
Swaps
Caps
Back-up generation
Load curtailment
Monitoring & reporting
Online monitoring
Monthly report
Process control
Environmental cost reduction
Purchase own RECs
Own small-scale solar
Own large-scale solar
Offtake agreement

Once we exhaust the ideas in the room for finding cost reductions in each of the main branches, we need to distill them down into the top one to three tactics that will give us the greatest return for the smallest investment in capital and resources in that branch. Here is an example of a simplified output from the brainstorming and selection process.

	NOW	FUTURE	HOW
Efficiency	100 kWh/unit	95 kWh/t	1. Energy-efficiency audit
Electricity unit cost	$90/MWh	$70/MWh	1. Wholesale market strategy 2. Research offtake agreement opportunity 3. Competitive tender
Environmental costs	$144,834	$36,209	1. Investigate the business case for large-scale solar 2. Investigate the business case for small-scale solar 3. Investigate offtake agreement opportunity
Network costs	$937,166	$656,016	1. Investigate business case for power factor correction 2. Identify opportunities to reduce peak demand 3. Review tariff change opportunities
Metering costs	$18,000	$6,000	1. Change meter agent
Maximum demand	1.2 MVA	1.0 MVA	Included in above
Power factor	0.833	0.98	Included in above
Annual consumption	10,000 MWh	9,000 MWh	1. Research monitoring and reporting options
Annual electricity cost	$2,000,000	$1,328,225	

6. Formulate ACTION

You then generate a series of actions and assign responsibility and deadlines for each of the actions. The actions should be in the SMART format: Specific, Measureable, Achievable, Relevant, Time-bound. The table on the following page shows some examples of SMART goals that you may set.

In this strategy formulation session we have identified a clear objective, identified the five strategies and supporting tactics that will enable us to achieve that objective, and fleshed out an action plan with timelines and accountabilities.

It's important for one of the team to project manage the strategic plan, hold people accountable for completing their actions and report on progress.

As the actions are delivered, more actions will be generated. For example, the energy-efficiency audit will identify a range of opportunities to reduce waste and increase efficiency. Some will be zero cost and easy to implement; some will require capital investment and will need to be justified and approved. Irrespective, the recommended actions have to be prioritised, approved, assigned and – most importantly – implemented.

Chapters 9 to 14 provide more detail in identifying the opportunities when developing this strategic plan.

	NOW	FUTURE	HOW (Tactics)	Actions	Who	When
Efficiency	100 kWh/unit	95 kWh/t	1. Energy efficiency audit	1. Engage Altus Energy to conduct an energy audit by 1 July	JS	1/7/2018
Electricity unit cost	$90/MWh	$70/MWh	1. Wholesale market strategy 2. Research offtake agreement opportunity 3. Competitive tender	1. Engage Altus Energy to model rewards and risks of wholesale market strategy by 1 July 2. Contact renewable energy proponents to investigate options for offtake agreements by 31 March 3. Engage Altus Energy to conduct a competitive electricity supply tender by 1 July 4. Evaluate competitive supply offers vs wholesale market strategy and decide on strategy by 31 July	JS JS JS JS	1/07/18 31/3/18 1/7/18 31/7/18
Environmental costs	$144,834	$36,209	1. Investigate the business case for large scale solar 2. Investigate the business case for small scale solar 3. Investigate offtake agreement opportunity			
Network costs	$937,166	$656,016	1. Investigate business case for power factor correction 2. Identify opportunities to reduce peak demand 3. Review tariff change opportunities			
Metering costs	$18,000	$6,000	1. Change meter agent			
Maximum demand	1.2 MVA	1.0 MVA	Included in above			
Power factor	0.833	0.98	Included in above			
Annual consumption	10,000 MWh	9,000 MWh	1. Research monitoring and reporting options			
Annual electricity cost	$2,000,000	$1,328,225				

CHAPTER 9

Step 2: Analyse your operations

In this step you analyse how you use your electricity, where you use it, when you use it, how you schedule it, and what the flexibilities are. This is important as the answers to those questions drive electricity costs.

Business size and electricity

The size of a business has a significant impact on how much energy is used, what energy costs are as a percentage of overall expenses, and how innovative the business will be in trying to implement an energy strategy.

Let's have a look …

Big, energy-intensive businesses

Different size businesses manage electricity procurement in different ways. The big, energy-intensive businesses often have a full-time energy procurement specialist who has often come to the role through one of two paths: either they were a procurement professional who worked their way up the procurement totem pole from stationery, to drives and bearings, to major contract management, and then to energy; or, the company identified that they needed an energy professional in the business and recruited an experienced energy professional from the supply side of the industry. Both types have a good understanding of how energy markets operate, have good connections in the industry, stay abreast of changes in the market through attending conferences and reading market reports and updates, and often get involved in end-user advocacy through one of the end-user advocacy groups. Energy costs often account for 15% to 30% of their total costs, and this is a high-profile area and position.

However, very few of these businesses employ an experienced operations professional who understands the actual operation of the assets, the bottlenecks, the under-utilised equipment, and the impact of turning equipment on and off. That's because they earn more money running operations than supporting the operations, and so don't want to do such a role.

Mid-size businesses

Mid-size businesses tend to have a procurement professional who looks after a wide range of categories, including energy. This professional usually does not have time to keep abreast of the market as they are always busy with other major procurement issues. Energy costs might account for 10% to 15% of their overall costs, and is still a significant cost item. This procurement person usually has limited understanding of the operations in the plant, and is concerned mainly with achieving the lowest unit cost for inputs to get the operations the best deal.

Smaller businesses

The smaller business usually has the business owner or senior business manager negotiate energy deals. Energy costs may be 5% to 10% of their total input costs. These people understand the operations intimately and are able to see the cost savings at the intersection of operational flexibility and flexibility in supply options. If they are the business owner they fully understand the impact of reward and risk on the money in their pocket and so seek to maximise reward, or reduced costs, and manage the risks. These business owners or senior managers usually have a much more entrepreneurial mindset and are willing to accept some quantified

risk in order to achieve much larger gains. They tend to be the most innovative.

The person responsible for energy in medium and large businesses is usually motivated toward ensuring security of supply and achieving the lowest unit price, without having to bother the demanding operations people too much. The operations people in those businesses don't understand the energy markets, and just want the procurement professional to get them the lowest cost and not be bothered with anything else.

It is often the more flexible, entrepreneurial owner or business manager that achieves the lowest cost of energy, even though energy costs may be a smaller percentage of their input costs.

Analysing the data

Your energy strategy is not going anywhere if you don't know how to collate and analyse your energy data. There are all sorts of different techniques you can use to do this. Different analysis techniques will reveal different information, so it's also important that you understand when to use each method.

Let's have a look ...

Box plots

The amount of data that is available at the operations level can be overwhelming when it comes to analysing electricity usage. For this reason, one of the best ways to analyse and represent the information and provide some context is to look at box plots for the data.

Box plots are a method of representing data that show the distribution of the data at a glance and not just over a time series average. Averages are not very useful when looking at electricity usage data as they obscure crucial information like the highs and the shape of the distribution.

Box plots are simple percentile plots where a box represents the 25th to the 75th percentile of the data, and "whiskers" either side indicate the expected variation in the upper and lower quartiles. For most box plots, the upper whisker extends to the highest data value within the upper limit range where the upper limit is 1.5 × the difference between the 25th and 75th percentile added to the

75th percentile. The lower whisker extends to the lowest data value within the equivalent lower limit range.

The median, where 50% of the data values are less than or equal to this number, is shown as a horizontal line across the box. We also show the average value in most of the charts as a solid dot with a line connecting them.

Data points that sit outside the extent of the whiskers are regarded as outliers and not within the expected distribution. They are due to special cause variation.

The following diagram shows a simple representation of a box plot for a set of data.

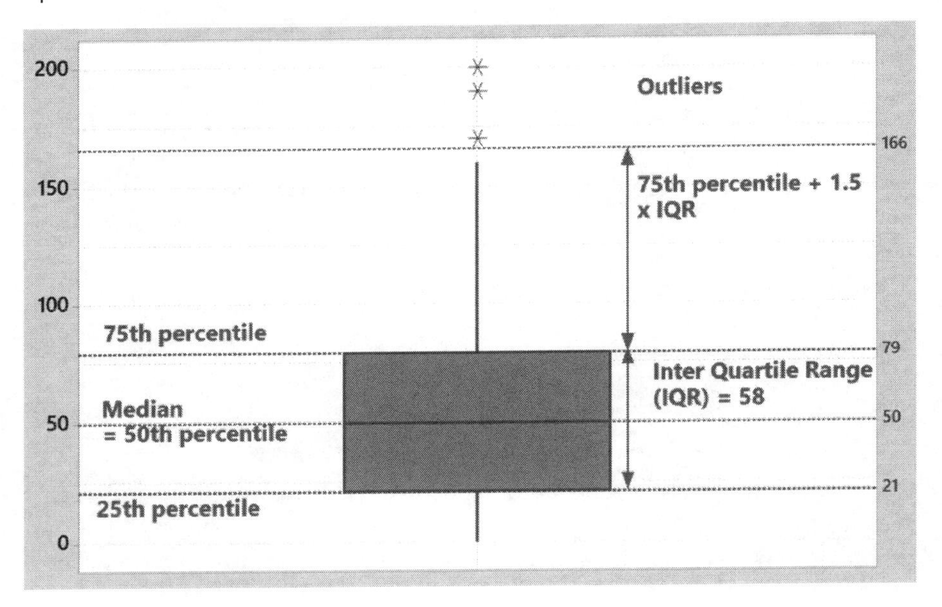

Interval data

Large electricity users will have meters that are read remotely and record the electricity consumption in 30-minute intervals, although sometimes this data will be in 15-minute intervals. This data is known as "interval data", and each customer can request this data from their retailer or sometimes the network provider.

It is common to look at 12 months of interval data, although I would usually analyse 24 months of data to get some more historical context. This is *a lot* of data. For 12 months this is 365 × 48 data points or 17,520 data points for each field. The data that we need is the time stamp, the kW and the kVA, which gives us 52,560 data

points. (The data set that is provided in the interval data actually has more data than this.) We then need to do some calculations and then present the data in a way that can be easily interpreted.

In Excel we will calculate the following from the data provided:

1 power factor
2 day of week
3 month of year.

We will also need to modify the time stamp to more easily allow us to graph the data. We then copy and paste this data into a statistical software package such as Minitab.

Once we have loaded this data into our statistical analysis software, we then need to slice and dice it a few different ways to look under the hood of the business in terms of how it uses electricity.

The way that we slice and dice the data and analyse it is we look at:

1 usage over the long-term trend
2 usage distribution by time of day (separating week days and weekends if appropriate)
3 usage by day of week
4 usage by month of year
5 drilling down to specific days
6 analysing power factor.

Run charts

A simple run chart of half-hour load data can reveal a lot about what is happening at a business in regards to electricity usage patterns. It shows the variability in load, it shows some seasonality, and it also shows step changes. Step changes in load are very important. Often there is a step change in the load upwards and the business has no visibility of that data. Electricity supply charges may increase by an amount that the accounts payable person doesn't notice and the responsible manager glosses over. However, when annualised, that amount becomes significant.

This first example shows two years of half-hour demand (kVA) data for a chemical manufacturing plant that runs on a 24 hour, 7-day week operation. This business is able to sell everything that it produces and so its profitability is based on maximising run time for as much as the year as possible, with short monthly stops for maintenance.

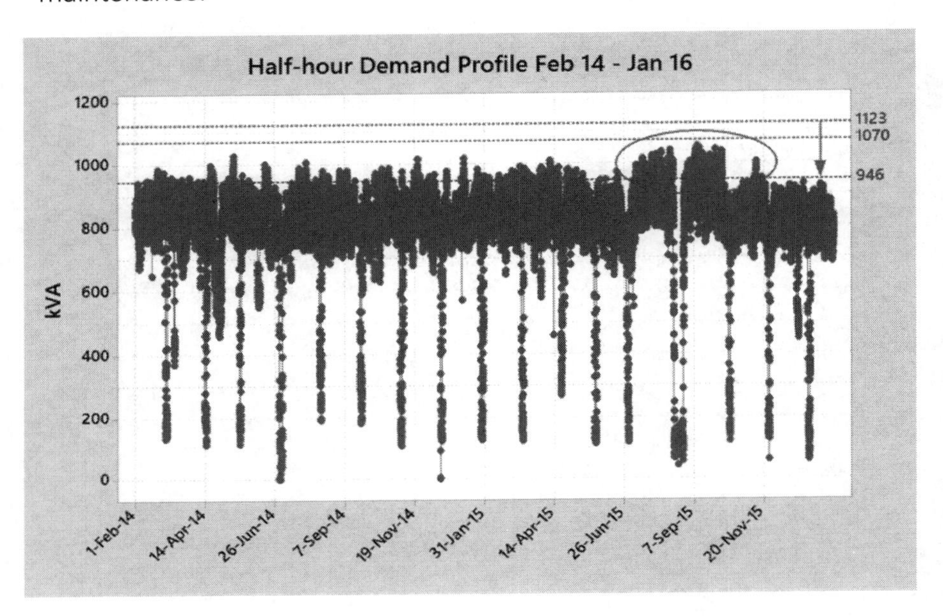

When I plotted a run chart of demand for this South Australian client I identified a large step up in demand in July and then a step down again a few months later. The step down, if sustainable, would allow the client to apply to have their network demand charges reduced by $16,623/year. And that's just from looking at one chart!

The cause of the increase in demand was solenoid valves leaking compressed air. This required the second of two large air compressors to be run, pushing up demand. The leakage was known about, recognised, and put on the maintenance plan to repair, but they were oblivious to the potential electricity cost impact due to the increase in agreed demand.

As it happened, the client was running below their fixed network peak demand level setting of 1,070 kVA and their anytime, or off-peak, demand level of 1,123 kVA so the step up would not have incurred additional cost. However, after making the request to reduce the agreed demand, any subsequent large air leaks causing

the second compressor to kick in would have increased their monthly and annual charges without the client even being aware.

Time-of-day data

Run charts are very useful as a first step, but looking at usage distributions at different times of day yields other interesting information that can lead to reduced costs. The following box plot chart shows an example of where a time-of-day pattern was revealed with analysis that led to a cost reduction.

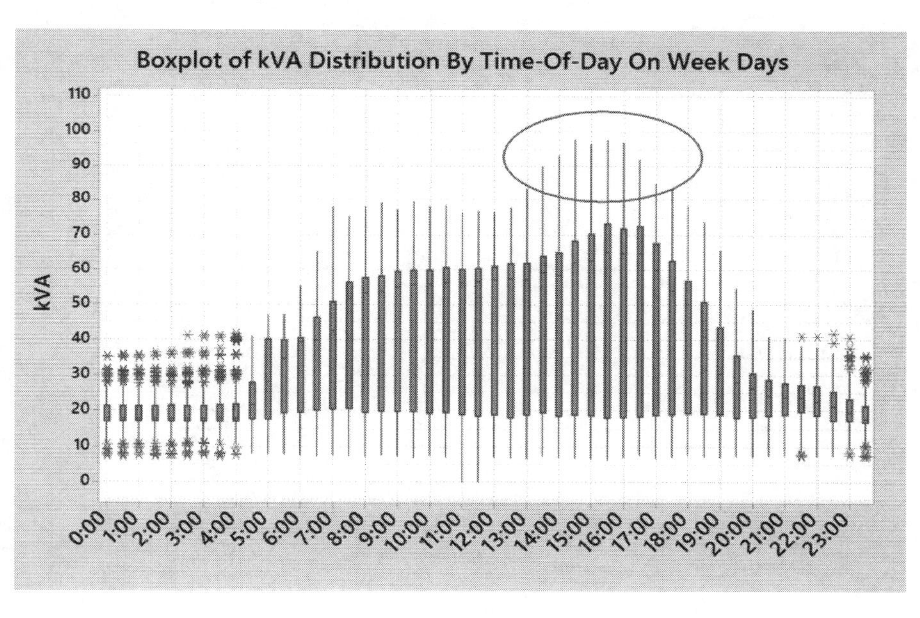

In this case the load was quite modest, but the time-of-day information revealed a cost-saving initiative that was completely unknown. The site is a warehouse and uses very little electricity other than lighting and air conditioning. Every weekday in the late afternoon the demand increased. This load pattern was repeatable irrespective of the season. What was even more unusual is that the increase in demand also coincided with when the employees finished work. When the cause of this increase was investigated it was found that it was the recharging of the electric forklift batteries. All the forklifts were being connected up at the same time at the end of the shift – during peak price periods. The solution to this was to install low-cost timers that delayed charging until the off-peak period.

Can you see how even a very simple analysis of the data and understanding of the electricity market can result in significant savings for a business?

Day-of-week data

Day-of-week data can reveal other patterns that can lead to reduced costs. The following chart shows how day-of-week analysis can highlight what is driving annual demand costs.

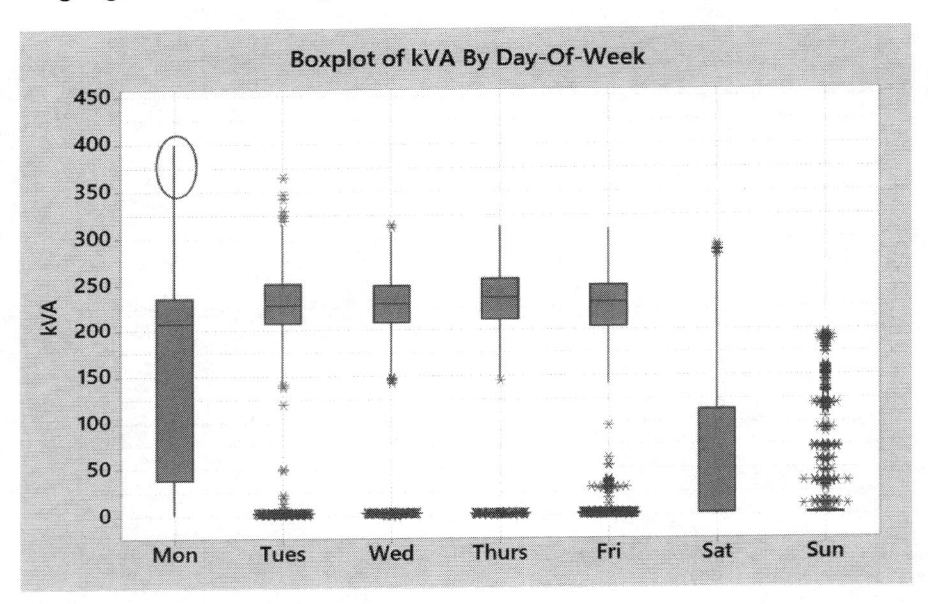

This chart shows that Mondays have a much higher maximum demand than every other day, but the distribution of half-hourly demand was less than any other week day. The factory operates generally five days a week, 24 hours per day, with some occasional market demand catch up on Saturdays.

It was quickly realised that the higher demand on Mondays, or Tuesdays if Monday was a public holiday, was likely driven by starting the factory up after being closed for the weekend. This is evident in the next chart that shows a very high demand between 4:30 am and 7:30 am, followed by a drop in demand throughout the morning.

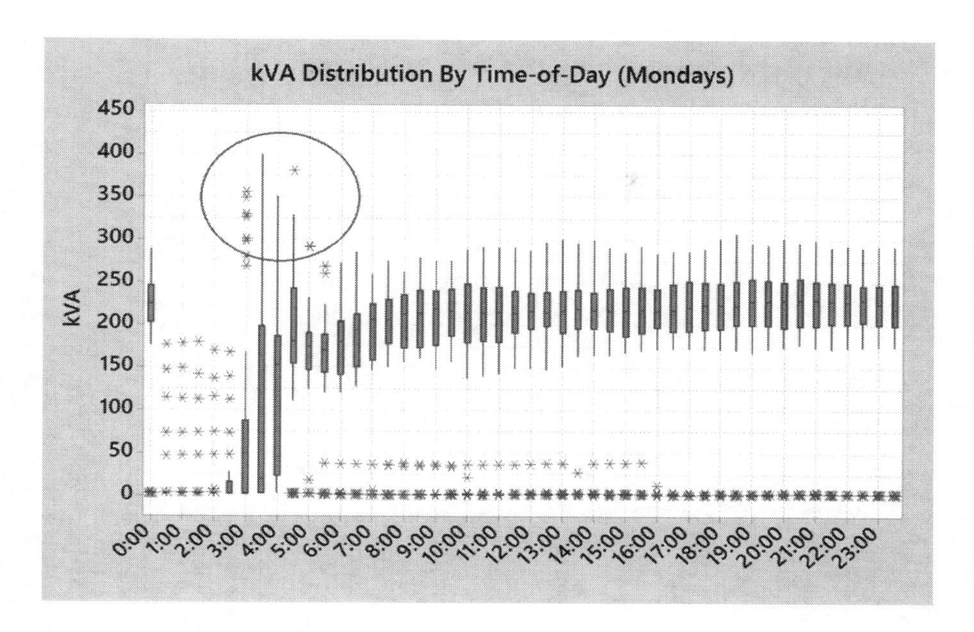

In this example the "additional demand" charges were being driven by Mondays. "Additional demand" is the period outside of the "peak demand" period and so has a lower cost.

Each Monday morning an employee came in early to start up all of the equipment and heaters after the plant had been off for the weekend. Even though this time was during the off-peak period, it pushed the off-peak demand well above the peak agreed demand and resulted in higher additional demand charges. The additional demand charges are not as high as the agreed demand charges. In this case, moderating the plant start up resulted in savings of $5,000 a year, which is substantial for a smaller operation.

Another important point is that if the start up of the plant was delayed until the peak period after 7:00 am then this would push up the more expensive agreed demand charges. So they were doing the right thing in regards to the timing of the start up of the plant.

Month-of-year data

Another pattern analysis that is interesting is month-of-year analysis. Similar to the run chart showing all of the data, this chart simplifies the representation of the data and shows the distribution over each month.

Month of year data is important to analyse as it can show longer term trends in electricity usage and demand, seasonal patterns and one-off or repeating problems that are impacting costs.

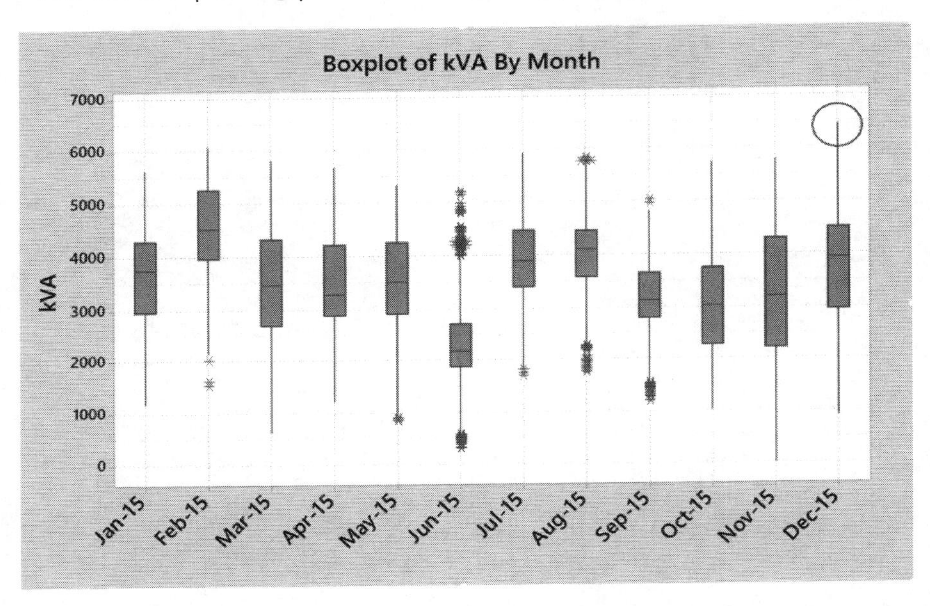

This relatively large factory had a significant spike in demand in December 2015, even though the median load was at a normal level. At a glance we can see that there was an issue in December. When December was investigated more closely, the culprit was found to be that the power factor correction capacitors had failed, pushing the demand (kVA) significantly higher for the same load (kW). This failure caused the agreed demand charges in the network tariff to be ratcheted up by more than $60,000 per year.

In order to identify the root cause we had needed to drill down into the specific day to see what happened throughout that day, and then examine the operating logs to see what happened in the operations to cause that spike.

We will examine power factor in more detail in Step 5 Reduce Your Network Costs as it is one of the important electricity characteristics of your business.

The electricity usage data that is used in this book has been obtained by interval meter data for the supply point. It is also possible – and sometimes very important – to monitor actual load at the sub-board level within a site. This can help to show what individual pieces of equipment or groups of equipment are running at any one time. This information is particularly useful in determining the actual run time of equipment loaded and unloaded, and so is an important input into load curtailment schedules that we will discuss in Step 6 of the framework: Reduce your energy unit costs.

CHAPTER 10

Step 3: Understand the electricity market

It's very important to have a good understanding of the workings of the electricity market when developing your electricity purchasing strategy. This allows you to make confident, well-informed decisions about the right time and the wrong time to go to the market for fixed retail pricing, whether retail price offers reflect good, fair or poor value, and what the opportunities and risks would be of adopting a wholesale market pricing strategy. You can gain an understanding of the market by analysing historical prices, and by putting today's pricing in the context of a larger timeframe.

If electricity is a high proportion of input costs for your business, it's very important that there is someone in the business, or an adviser, who understands the electricity market price outcomes, current trends and regulations, and reforms that may be around the corner. Understanding the market means you can make the connections between your business's unique circumstances and methods of operation and the many opportunities to reduce costs. Being aware of the current trends and reforms means you can stay ahead of the game when those changes occur.

As we've seen, there is a correlation between current spot market prices and current retail prices. As the spot market becomes more volatile and prices rise, so does retail pricing. If you were looking to go to the market for retail price offers, you should first understand what is happening in the market to ensure you are not requesting fixed-price offers after a recent period of high and volatile prices.

An overview of price drivers

So what drives prices? Is it simply the case that when demand goes up, prices go up, and when demand drops, prices fall?

Electricity is a unique commodity in that it is generated and consumed instantaneously. Excess generation cannot be easily and cost-effectively stored for use at a later time when demand is high, although that could be about to change on a modest scale with new battery technology. This results in price behaviour that exhibits time-of-day and seasonal demand characteristics, as well as random spikes due to disruptions in supply. In addition, generators can affect price through bidding strategies.

The spot price is determined by the intersection of supply and demand curves. The supply curve is established by generators making a series of short-term bids at different prices – according to each generator's bidding strategy – that are arranged in a supply bid stack. The actual demand then determines both the spot price and the generation that is scheduled. All generators bidding at or below the spot price will earn the spot price for the energy that they supply according to the quantity scheduled.

For our purposes, we will use what is called "scheduled demand" as the measurement of market demand. Scheduled demand is the generation required to meet regional demand less non-scheduled and exempt generation. Non-scheduled generation is generally some small-scale wind and solar generation and small local generation on a local consumer site.

There is not a simple relationship between demand and price though. The generators are changing their bids every five-minute interval for a wide range of reasons, including being strategic to maximise their profits. High prices are more likely to occur during periods of high load, but the correlation turns out not to be strong.

Long-term prices will be affected by the growth in both supply and demand, and, in particular, the ability of growth in supply to meet peak demand across the NEM. Another major factor in the changes in long-term prices is changes in the long-term generation mix. As older coal-fired power generators are closed, such as

the Port Augusta Northern Power Station and the Hazelwood Power Station, the market may become more reliant on gas-fired generation. This means that price will be affected by short-term and long-term gas prices and the mix of gas generation technology.

The factors affecting electricity prices in the short to medium term are:

- demand – net of wind
- supply – the bid stack curve
- supply shocks – network and generators shocks.

We will focus on using South Australian data to explain the factors driving prices, as South Australia has experienced some extremes in recent times and the examples stand out. However, the same principles apply to other regions.

Demand

As a general principle, the higher the demand the higher the price should be. If there was a static bid supply stack then that would be the case, however, as we can see in the next graph, there is a very weak linear correlation between price and demand.

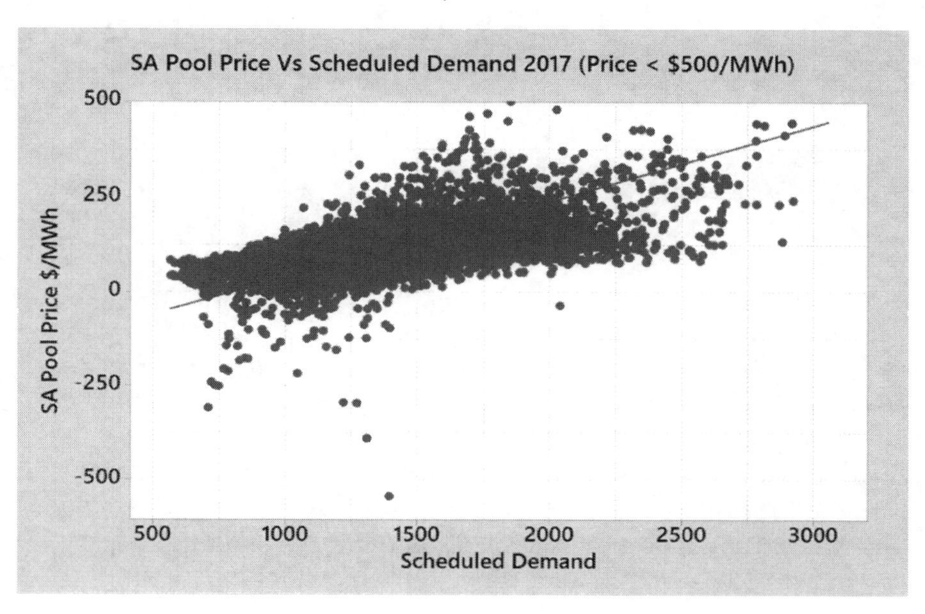

This scatterplot chart shows that there must be factors that impact price other than demand, although demand has an influence.

Let's look at the drivers of demand.

Time of day

Demand cycles throughout a 24-hour period as businesses, schools and services start up in the morning and then shut down in the early evening. Households use more electricity in the early mornings as people wake up, and early evenings as people arrive home.

The chart following shows the actual historical demand fluctuation in South Australia throughout working days in the summers of 2014–15 and 2015–16.

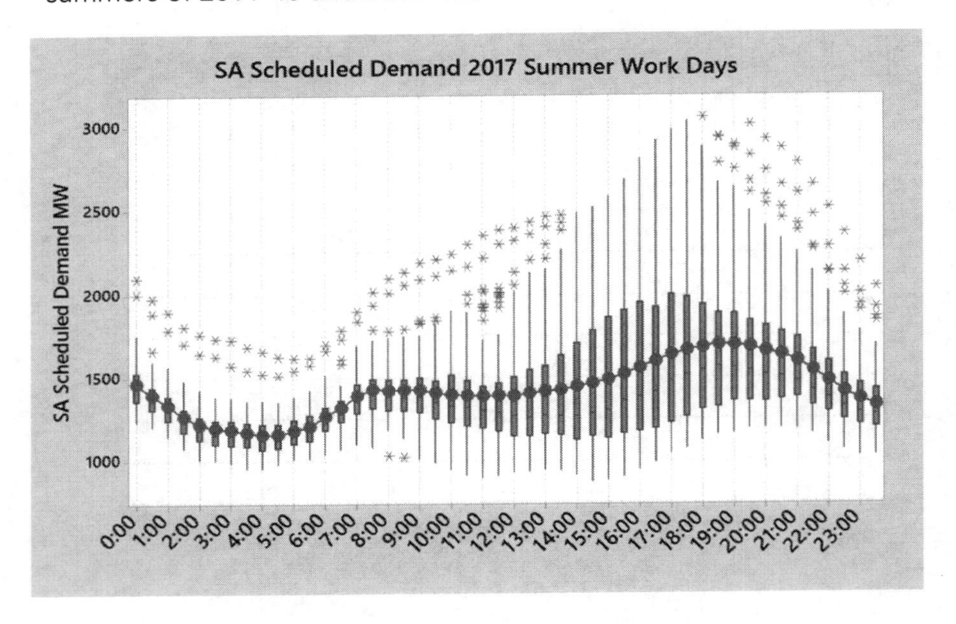

This box plot shows the distribution of data, and that distribution is very wide throughout summer day times and early evenings. South Australia suffers through some very hot summer periods, but it can also enjoy quite mild summer temperatures.

The amount of wind generation also has an impact on the reported demand as some wind generation is non-scheduled by the market operator and effectively reduces the scheduled demand by the amount of non-scheduled generation. However, the amount of non-scheduled wind generation has reduced recently and so has only a small impact on this analysis. Small-scale solar PV also effectively reduces the scheduled demand.

There are some key characteristics of the summer working-day pattern. Demand drops to a consistently low level overnight and starts to increase at around 5 am. Demand continues to increase as people wake up and turn on electrical appliances, and then steadies out at around 8 am when people head off to work, the kids to school, and businesses have finished starting up for the day. While there is obviously a lot of variation between days depending on the daily temperature, average demand remains relatively flat until the afternoon, when it gradually increases into the early evening. Demand reaches its peak in the early evening between 5 pm and 7:30 pm, and it can go to very high levels, stretching the supply capacity. After around 7:30 pm it begins to sharply drop.

So how does that play out in electricity price patterns?

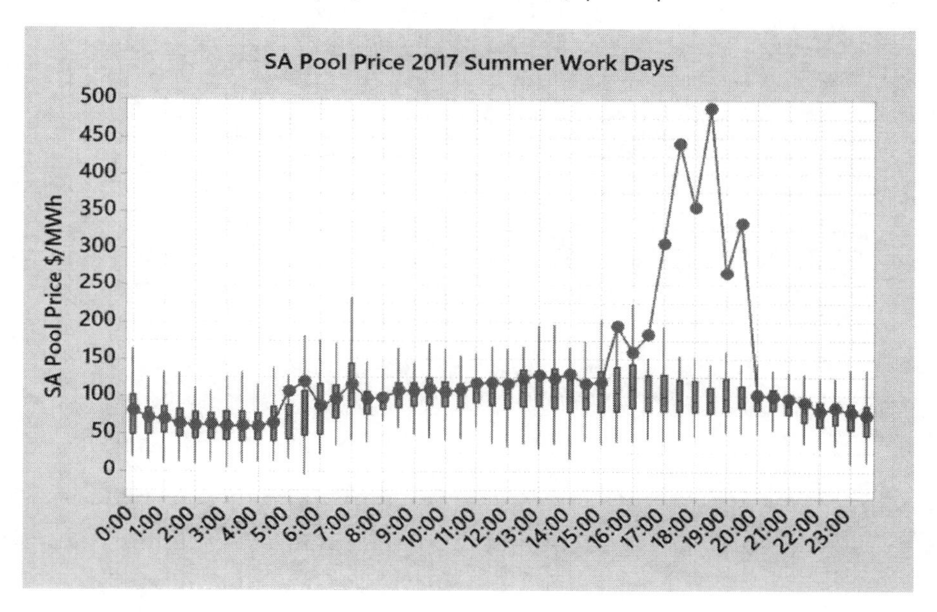

The spot price charts exclude outliers as they distort the view of the bulk of the data. However, those price spikes are evident when the average trend line moves above the box plot median and sometimes well above the box whiskers. In this summer chart, prices follow a similar trend to demand, with low overnight prices, and increasing from around 5:00 am. Price spikes in the late afternoon due to the very high demand on very hot days have driven the average price well above the median prices, and there is a steady decline in the average and median prices from the early evening.

If you were exposed to wholesale spot prices and you had operational flexibility, you could schedule your major loads to avoid late afternoons in summer to achieve lower prices, especially on very hot days.

Let's now look at a winter demand pattern for South Australia.

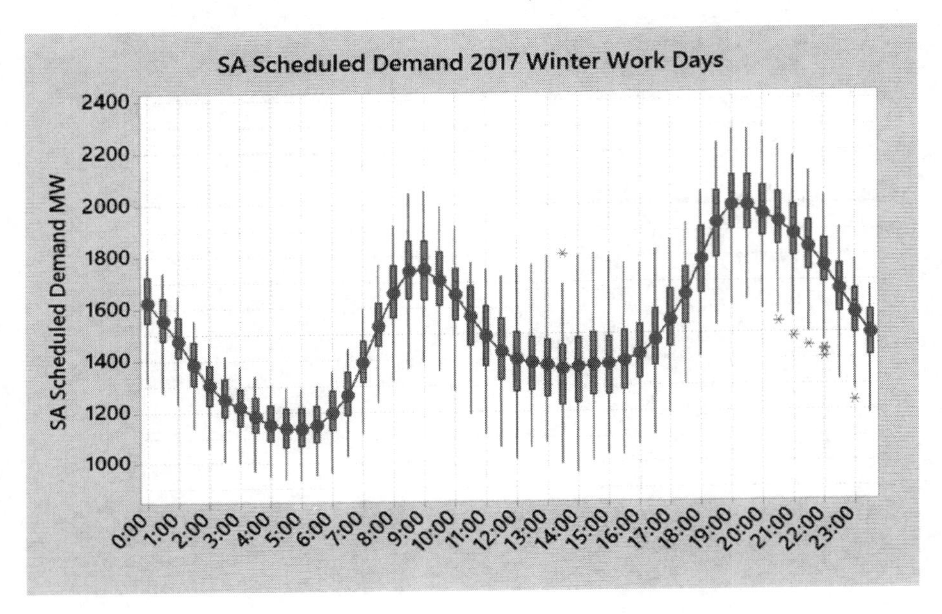

The winter demand data is far more predictable than the summer period as there is less fluctuation in temperature. In winter there is a significant amount of electricity used to heat up houses when people wake up in the morning and then arrive home after work in the early evening.

On average, this is reflected in the pricing, as shown in the next graph.

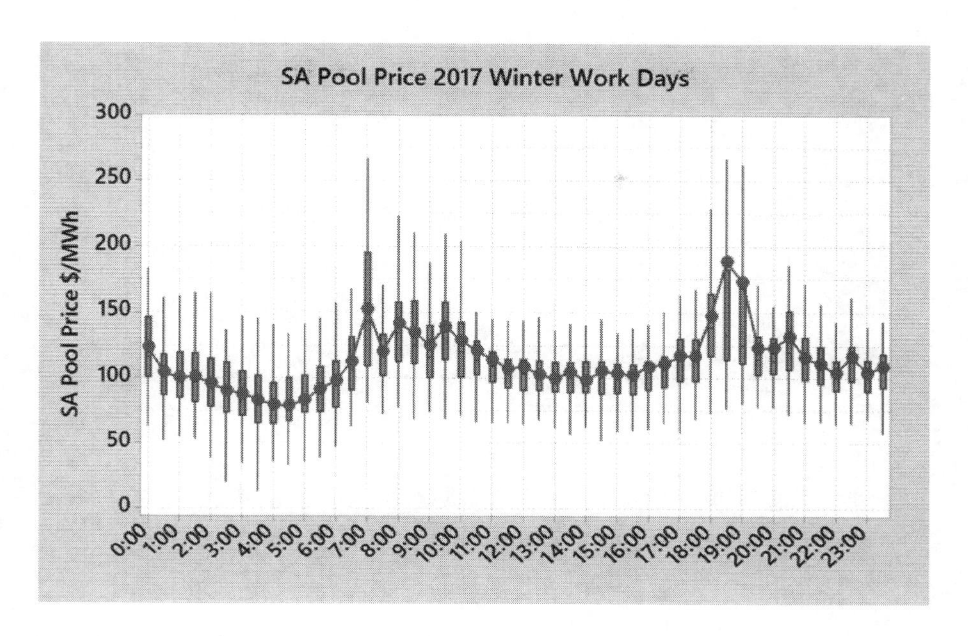

Once again we can see that prices drop to lower levels overnight and start to increase after 6:00 am, with a very predictable increase in the half-hour period from 6:30 am to 7:00 am. Prices drop away in the late morning and then increase sharply again in the early evening, with some price spikes dragging the average up in the period 5:30 pm to 7:00 pm. Prices then fall quite quickly during the evening as per the demand falling away. These patterns are predictable and the high prices may be avoidable for some businesses.

Day of week

Businesses, schools and services activities account for a large proportion of demand with energy-intensive machinery and lighting, heating and cooling. A large amount of this electricity is used during week days.

The following two graphs shows the SA demand distribution by day of the week for both summer and winter.

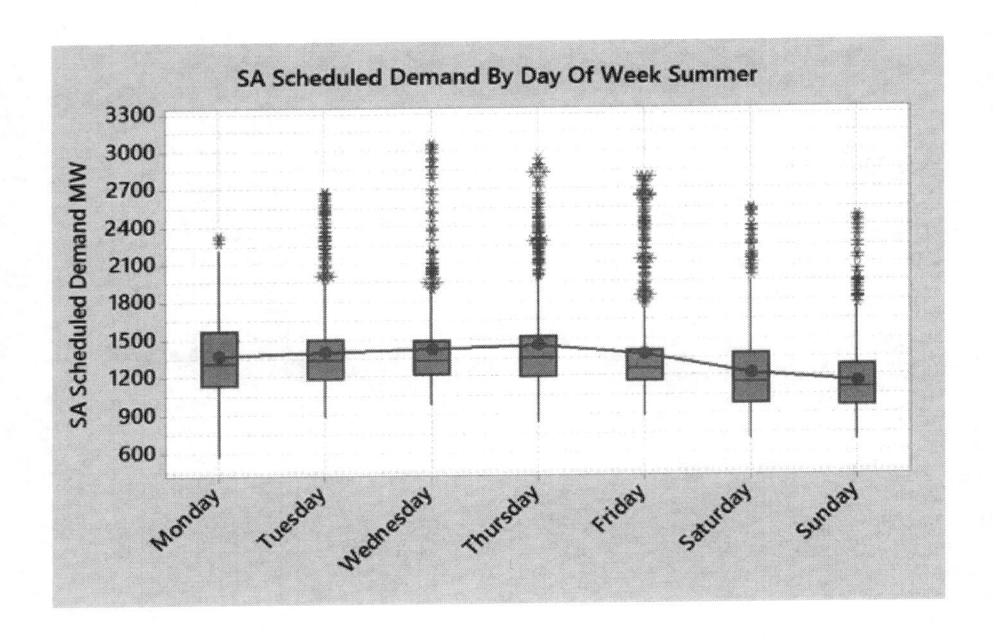

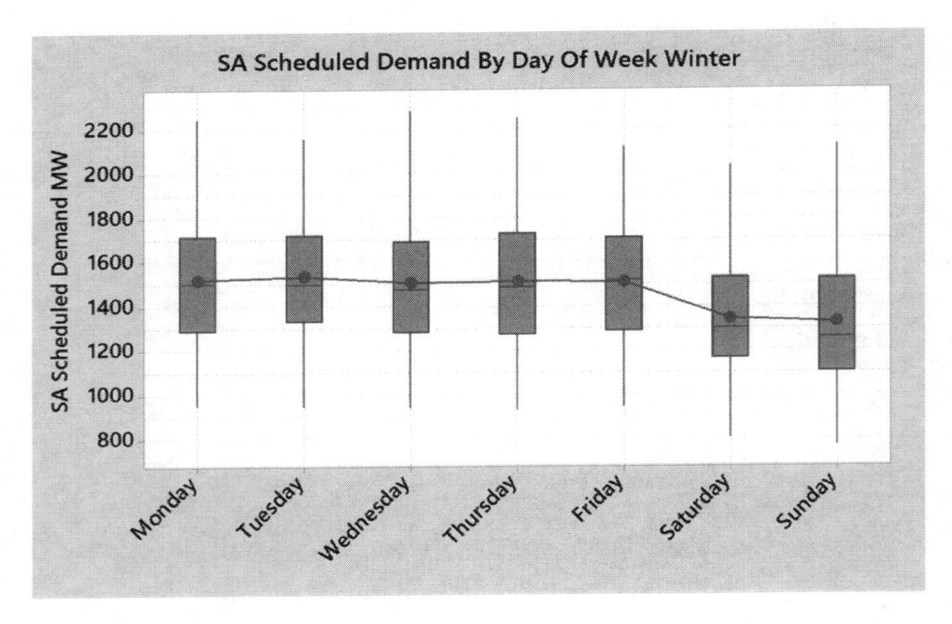

It can be clearly seen that demand is much lower on weekends than during week days in both summer and winter, as would be expected. The spikes in demand due to very hot weather are evident in the summer graph and absent in the winter graph. This is generally how pricing plays out as well.

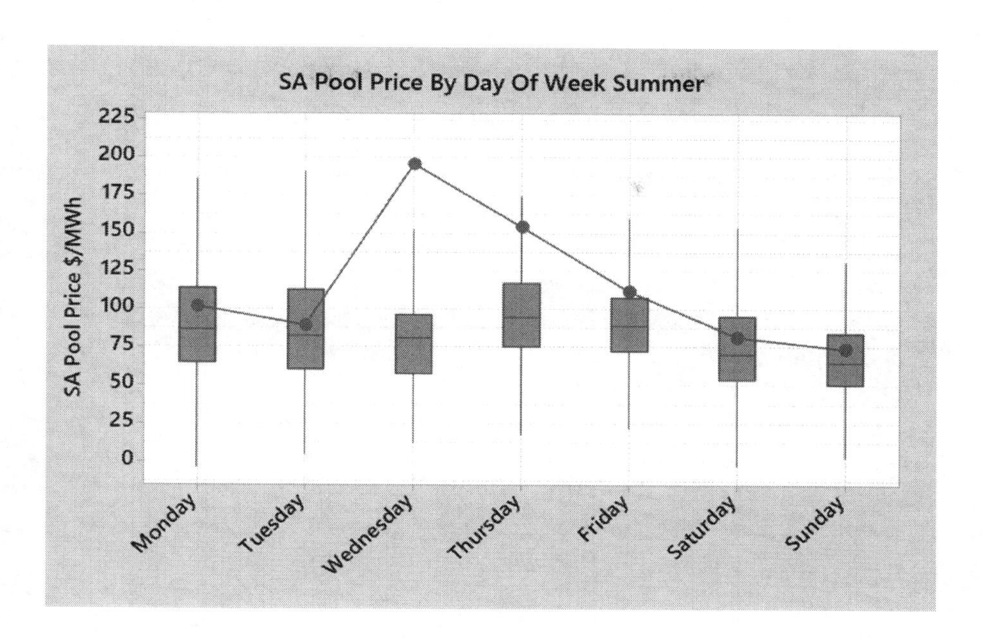

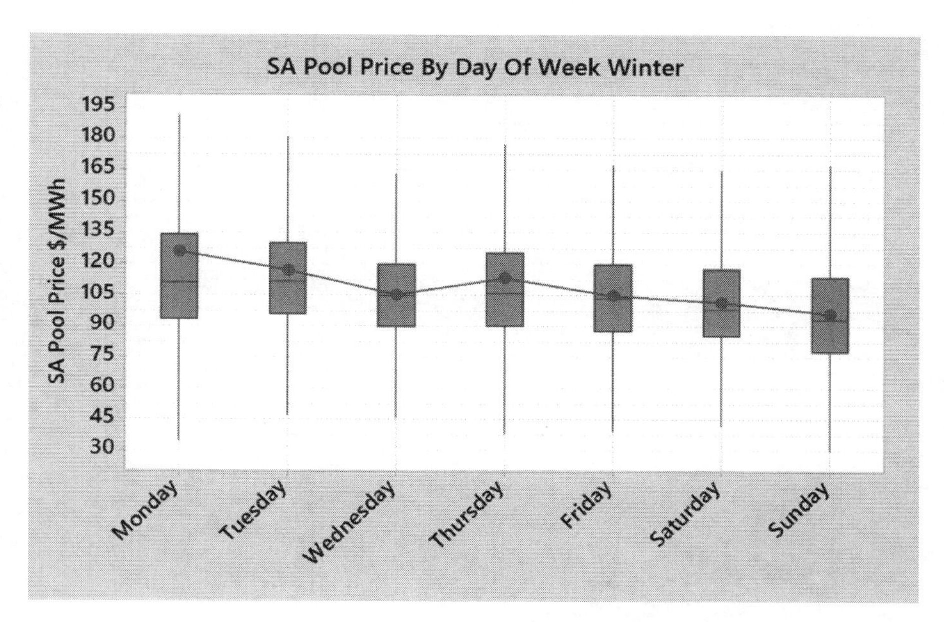

We can see that spot prices are lower on weekends, as would be expected. This lines up with fixed retail pricing, where the off-peak period is generally overnight and on weekends.

A business that has flexibility in when they run their major loads would know that they would save significant costs by scheduling to run predominantly at night time and weekends. Irrigators and water utilities, for example, can plan to maximise the operation of their pumps at night and on weekends.

Wind generation output

South Australia has a large and increasing reliance on wind generation. Wind generation is automatically dispatched at a floor price of –$1,000/MWh. The gap between wind output and actual demand, net of solar, is filled by other generation, mainly gas-fired. So it is the gas-fired generation that will set the price, provided that demand, net of solar, exceeds wind generation. The larger, more efficient, combined-cycle gas turbine station at Pelican Point, and the gas thermal stations at Torrens Island, are the next lowest cost generators. To ensure that they are not continually turning boilers on and off, they will bid prices in to the market at very low or even negative prices to ensure that they are dispatched ahead of other generators. This can result in very low or even negative pool prices. End users can actually be paid to use electricity, within a given half-hour period.

So the level of wind generation should have an impact on spot prices. As we can see from the next graph, the level of wind generation is highly variable, and ranges from as low as zero to as high as 1,594 MW (note that higher wind output levels have been achieved in 2017). Actual metered output averaged almost 557 MW over the 2017 period and the median was 485 MW.

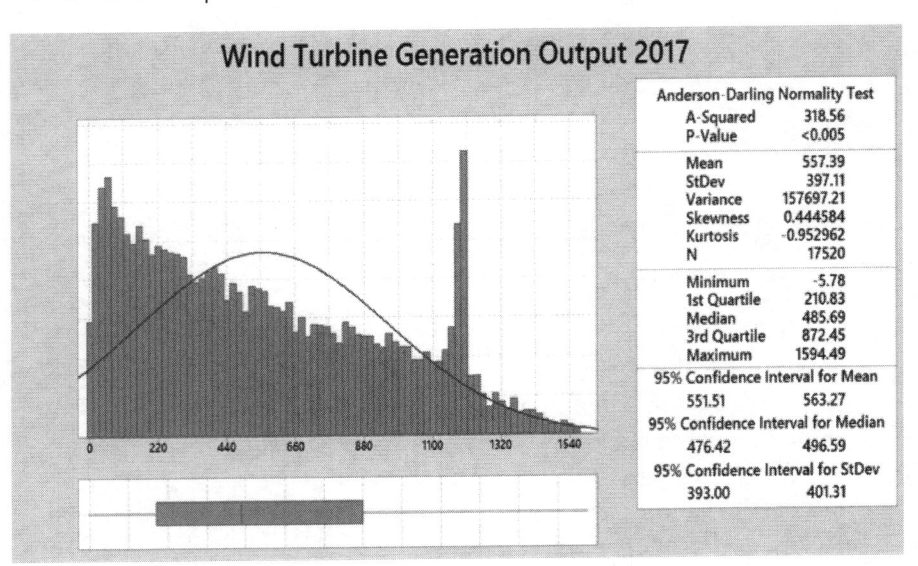

The level of wind generation compared with the scheduled demand should be expected to have a large influence on spot prices. The graph following plots the distribution of "scheduled demand minus wind" and shows the data distribution. The correlation coefficient between price and "scheduled demand minus wind" generation is still quite weak due to the other factors that influence price, such as the type of generation required to be dispatched to meet the last incremental demand.

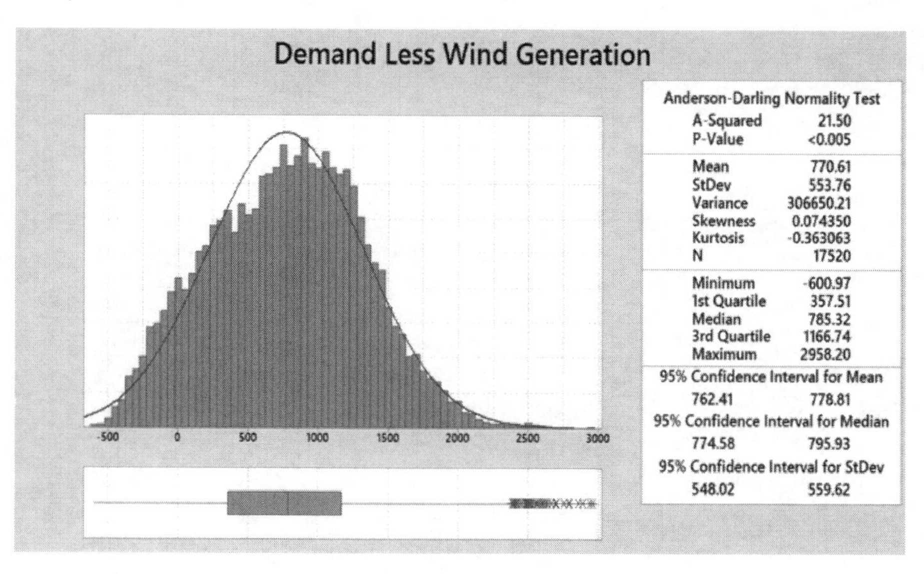

If we categorise "scheduled demand minus wind" generation into low, medium, high and very high "buckets" based on the quartiles of "scheduled demand minus wind" shown in the graph (25% 358 MW, 50% 785 MW, 75% 1,167 MW), we can examine the price distribution during these different scenarios.

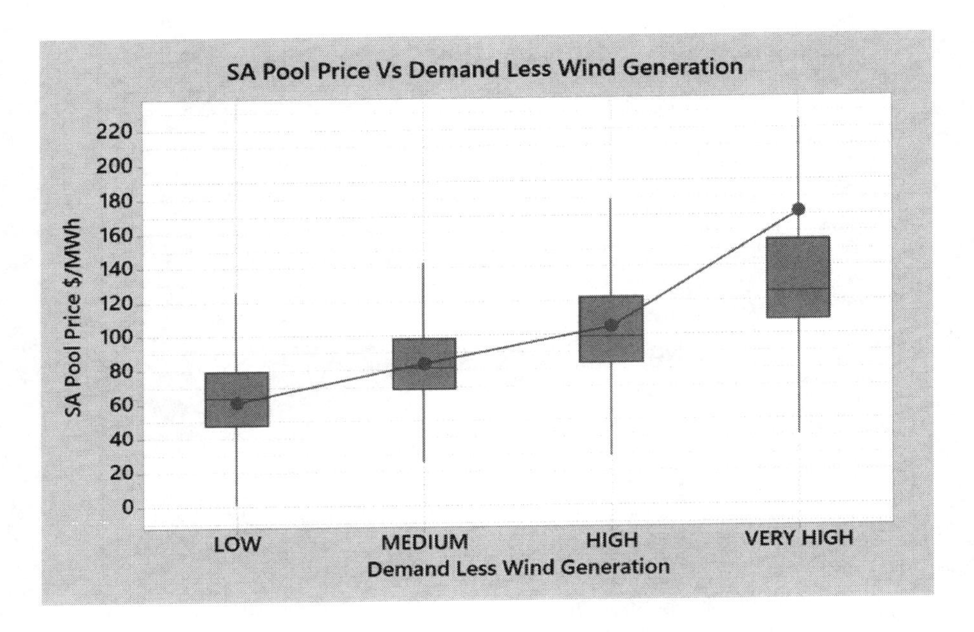

The impact of wind is quite clear in the previous graph. There is a clear step up in the distribution of prices for the different low, medium, high and very high "scheduled demand minus wind" scenarios. What's more, the difference between the averages – the symbols joined by the lines – and the median – the horizontal lines – increases under each higher scenario. This indicates that there is a greater frequency and/or level of price spikes during higher demand net of wind scenarios.

What can we conclude from this? When the wind is blowing and demand is low, prices are very low and very stable. When the wind is not blowing and demand is high, prices are very high and volatile. This demonstrates the importance of wind generation in determining SA spot prices.

Inelasticity of demand

Price elasticity is the percentage change in quantity demanded by the percentage change in price. Price elasticity is more "inelastic" as it moves towards zero. Many end users have fixed price contracts and therefore no incentive to react to fluctuations in the wholesale price of electricity. This results in price inelasticity. In the presence of market power, suppliers have the ability to set prices above the cost of the last unit produced. This ability increases with increasing

price inelasticity. Price spikes are more likely to occur when the expected demand is high and the level of market power is at its greatest.

Demand-side response increases elasticity and reduces the incidence of short-term market power. It sends a signal to the market about the value of electricity to the end user.

Customer exposure to short-term price signals – the spot market – is a small proportion of the total market. As the exposure grows there will be greater customer or demand-side response to high prices. The corollary is that as more large customers take on pool price exposure, the greater will be the level of demand response to price and there will be less market power on the supply side.

Having said that, one senior retailer executive pointed out to me that the greater the amount of load that is exposed to the spot market, the greater the level of uncontracted generator capacity. Generators tend to price contracted generation capacity at lower prices to ensure that they are dispatched and increase bid prices for uncontracted capacity in order to maximise profit.

Supply

Regional market generator mix (type and capacity)

The type of generator is a major determinant of the prices it will bid to supply. Generators bid at prices at which they will supply a block of electricity. Storage to smooth out changes in production capacity availability or changes in demand is not yet practically or economically possible, except in the case of hydroelectric power. Therefore the price is set by the generator of the last kilowatt of electricity required to meet the instantaneous demand.

The sum of all the generators' bids determines the bid supply stack curve. The bid supply stack is a curve of increasing generation capacity available on the horizontal axis and increasing price on the vertical axis. Electricity market bid supply stacks are generally long and flat on the left-hand side and rising to near vertical on the right-hand side.

Bid stack example

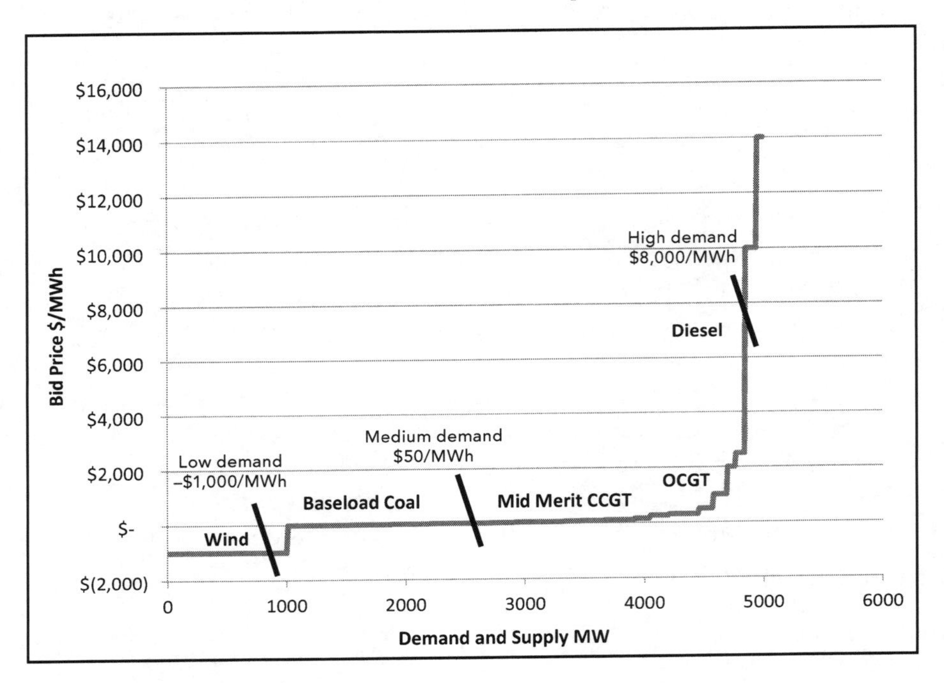

Coal-fired generators are large, capital-intensive plants with high fixed operating costs and low variable costs, and long run-down and start-up times. These generators make money by operating as often as possible to cover their fixed costs, and therefore set bid prices to ensure they sell as much of their capacity as possible without having to continually ramp up and down their output or turn boilers on and off. They may operate boilers at less than full capacity to meet their contracted demand and then bid the remaining capacity in at higher prices to capture some upside on non-contracted requirements.

Coal-fired generators show up on the supply stack curve as the long horizontal line at lower demand levels. Their main variable cost is the cost of coal, which will vary based on proximity to the mine, transport costs, mining costs, whether they own their own mine, and coal quality. If they don't own their own mine then they may be exposed to fluctuations in world coal prices, as usually coal prices are linked to a global trading index.

In theory coal generators will bid in prices that are at or just slightly above their variable costs, but quite often they may bid in

lower costs to ensure that their desired capacity is scheduled and dispatched ahead of other coal-fired generators.

Gas-fired stations' main variable is the cost of gas. This will depend on their upstream contracted costs for gas, gas pipeline transport costs, exposure to short-term gas prices, and the efficiency of the plant in terms of gas energy consumed to produce a unit of electricity. This will depend on the technology: thermal steam, combined-cycle gas turbines, open-cycle gas turbines and gas engines.

Diesel-fired generators' main variable cost is the cost of diesel and the delivery costs of diesel.

On the right side of the bid supply stack are peaking units that have high variable (fuel) costs and low fixed operating costs and low capital cost. These generators have very short start-up times and run-down times, and only run in periods of high spot price when demand exceeds the capacity of the other generator types. Consequently these plants have low utilisation. These types of generators are typically gas-fired turbines or oil/diesel-fired units, and the size of the plant, technology type and fuel costs will result in different bid strategies. This results in a fairly smooth, convex supply bid stack with prices rising to very high levels quickly in periods of high demand.

In between the two extremes of the supply bid stack are "mid-merit" units that will have different fuel costs depending on the region they operate in, and will have capital costs and fixed costs between those of the peaking units and base-load plants. Typical such plants are combined-cycle gas turbines and gas thermal stations. Torrens Island and Pelican Point are examples in South Australia.

Interconnectors with other regions bring in more supply capacity, although this is constrained by the capacity of the interconnector. The generators in other regions will affect the shape of the bid supply stack, with supply depending on overall market demand factors across all regions.

Generator bidding strategies

Generators are profit-making entities. Their bidding strategies are designed to maximise profit and minimise risk. Their bids change throughout the day to reflect this.

Generator bidding strategies will largely be determined by generator capacity and type, but also by profit-maximising considerations. The strategy will change depending on the return requirements of each generator and opportunistic profit-maximising strategies. Generators may have capacity they can make available at the marginal cost of production, but instead choose to bid that capacity at a higher price and are happy for that capacity not to be dispatched. Often the strategy will be to bid in prices to maximise the probability of being dispatched to meet their contracted capacity, and then bid in higher prices to maximise the profit opportunity for non-contracted prices.

Open-cycle gas turbines and diesel generators often provide the underlying physical generation for price cap products, which we will discuss later. In brief, some market intermediaries sell retailers or consumers a contract to ensure that the buyer does not pay more than a cap price of, say, $300/MWh. They may only generate at or above the cap price.

Generator market power

In a perfect market where there is a demand-side response to prices, customers would signal to the market the value they place on electricity. However, in the national electricity market most end-use consumers are not paying the market price as they are on fixed contracts and therefore they do not send price signals to the market. The suppliers therefore determine the pricing mechanism. In the presence of competition, price bids will reflect the marginal cost of production as generators compete for volume. In periods of high demand when most of the generators are fully committed, the very few remaining generators with spare capacity can exercise market power and bid prices that are well above the marginal cost of production.

Far from being a market failure, the ability to exercise this market power at periods of high demand is critical to the investment in new generation plants. A high frequency of price spikes indicates that there is a shortfall in the availability of low-cost generation capacity and sends a longer term market investment signal.

The Council of Australian Governments' "Independent Review of Energy Market Directions" found that price spikes generally reflect tight supply–demand situations which confer temporary market power on some generators, but that as dispatch occurs on a five-minute basis and prices are known to the market very close to real time, the duration of this market power is limited.[3] The review also found that these price spikes play a valuable role in sending demand-side signals of supply scarcity as well as necessary signals for investment in peaking plants.

The high price spikes seen across the NEM in 2016 and early 2017, along with the retirement of old, inefficient coal-fired stations, may well be the price signals that the market was designed to send for investment in new base-load capacity. It may well be the case that there are some investment projects in the pipeline for low-carbon-emission-technology coal plants, combined-cycle gas turbines and even large-scale solar or thermal.

The problem with this market signal in the current climate is that there is future uncertainty about whether there will be a carbon tax or trading scheme, and what that would look like if it was introduced. Power companies are unwilling to invest in large base-load capacity without that certainty. The market is sending a signal that new investment in base load is required but the only investors that are listening are investing in solar and wind.

The current high prices are not a market failure. The market was designed to operate this way and send new investment signals. The lesson for business owners and managers who are developing an energy strategy for their business is to understand that this is part of the cycle of retirement of old, high-cost, inefficient power generation, to be replaced with new technology and more efficient plants.

Generator maintenance cycles

Generators require periodic – for example, annual – maintenance that may require outages of several weeks, or even months. During this time, supply capacity is withdrawn from the market. Although such outages will generally be scheduled for low seasonal demand periods, they will still significantly change the bid stack for the region. This may result in a general increase in prices during this period, or result in the market price being more susceptible to other supply interruptions.

---

3 Parer, 2002, pp 104–105.

Interconnector capacity with other regions

Interconnectors allow different regions to import and export power. The capacity of these interconnectors will largely determine price differences between regions. A state with relatively low interconnector capacity will be exposed to higher prices with higher localised demand, as there will be a constraint as to the amount of lower priced electricity from an adjacent region that can be transported to the local region.

Supply shocks

Generator failures

Occasionally generators suddenly trip (shut down) due to unplanned electrical or mechanical problems. This will have the effect of instantaneously removing a block of energy from the grid and producing a sudden spike in costs as higher cost, fast start-up generators are automatically dispatched to supply the electricity.

Interconnector failures

Failures of interconnectors between regions will result in restrictions of supply in the importing region and an increase in the capacity of low-price supply in the exporting region.

A good example is if the Victoria to South Australia Heywood interconnector fails in a period of high wind generation – South Australian prices will likely go negative. If there is low wind generation then prices could spike to extreme levels and there could be a possibility of blackouts.

Gas supply failures

On 1 January 2004 there was a major explosion and fire at the Moomba gas field. This resulted in the cessation of gas supplies from Moomba in the short term and gas supply restrictions in the medium term. Fortunately, and coincidentally, a second gas pipeline to SA from Victoria (SEAGas) was in the process of commissioning. Commissioning was brought forward and gas supply was maintained, but at higher gas charges as this gas was largely uncontracted. This increase in gas costs resulted in higher bid prices for electricity supply.

Supply and demand conditions in each state

All of the NEM regions are connected via interconnectors that transport electricity in both directions. High demand and prices in one region therefore flow into other regions as generators can bid into different regions. If demand and prices are very high in NSW, high prices can be experienced in both Queensland and Victoria as generators in those regions make blocks of electricity available for NSW and modify their bid offers for different energy bands in their own region.

In 2017 we have seen a very high correlation in prices across all regions. In the past, extreme prices in South Australia would have been largely contained to South Australia. In 2017 we have seen the market prices move much closer together across regions.

Historical context across each of the regions

Now that you understand the drivers of price within a region, we can examine each region in a historical context and examine current pricing levels against that backdrop. Again, this is valuable information for business owners and managers looking to develop an energy strategy for their business.

The following charts show the historical distribution of prices in a box plot format for each of the regions since the commencement of the NEM. The outliers have been excluded but the annual average has been shown for each calendar year. Note that the "price on carbon" was brought in on 1 July 2012 and then repealed two years later on 1 July 2014. There was a step up in prices roughly in line with the price on carbon in the second half of 2012 and then a step down in 2014 of a lesser magnitude than the carbon price.

Queensland

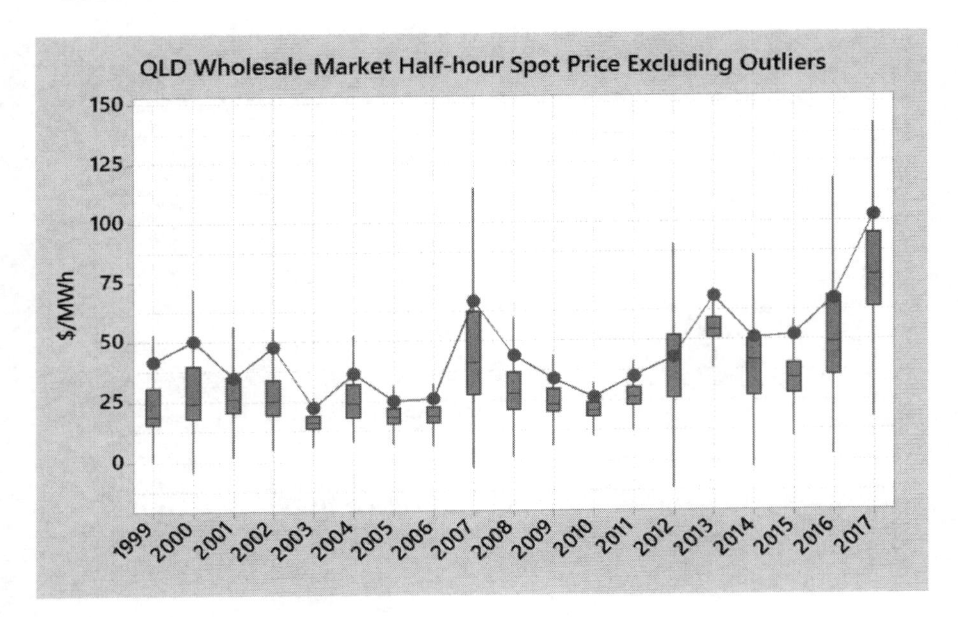

Queensland enjoyed very low spot prices from the commencement of the market up until the commencement of the carbon price, with the exception of the 2007 drought year that carried in to early 2008. Years 2012 to 2014 were impacted by the price on carbon. However, even when the price on carbon was repealed, wholesale spot market prices remained high, on average, due to the impact of summer price spikes dragging the average well above the median price.

The drought caused large coal-fired power stations that used fresh water as coolant to restrict their operations and so reduce supply. This impacted the bid stack and drove prices higher. Will Queensland be affected by drought again in the future? Yes. Will this impact spot market prices? Most likely yes. Will drought occur every year? Most likely no.

Can you use this historical information that includes the impact of droughts, major legislation changes and sudden supply and demand disruptions within the context of long-term median prices in determining whether it is the right time to go to the market for retail prices or whether your retail price offers are high, low or fair value? Absolutely yes.

For consumers, 2017 was the worst year on record, due to the very high first quarter prices that averaged $173/MWh. That was the worst quarter in history. Average prices dropped below $90/MWh for the remaining three quarters of the year. The first quarter is usually much higher due to demand, however that quarter was impacted by the closure of the Hazelwood Station in Victoria. The question is, based on the historical context and your market intelligence of what is likely to impact the market in the coming years, what are the expectations for prices in 2018 and beyond?

Looking at the previous chart, would you accept a retail price offer for the 2018 to 2020 period that was equivalent to a $100/MWh flat price? Many businesses do. They believe that accepting these very high fixed retail prices will protect them from the volatility and risk of future price fluctuations. Would you accept an offer of $90/MWh for the next three years? Well, that depends on your tolerance for risk and ability to absorb month-to-month variability in input costs. Would you accept a fixed price of $60/MWh for three years? I would. It all depends on your strategic energy objective, your ability to curtail load, and your tolerance for uncertainty – and your understanding of the market.

New South Wales

Let's have a look at NSW.

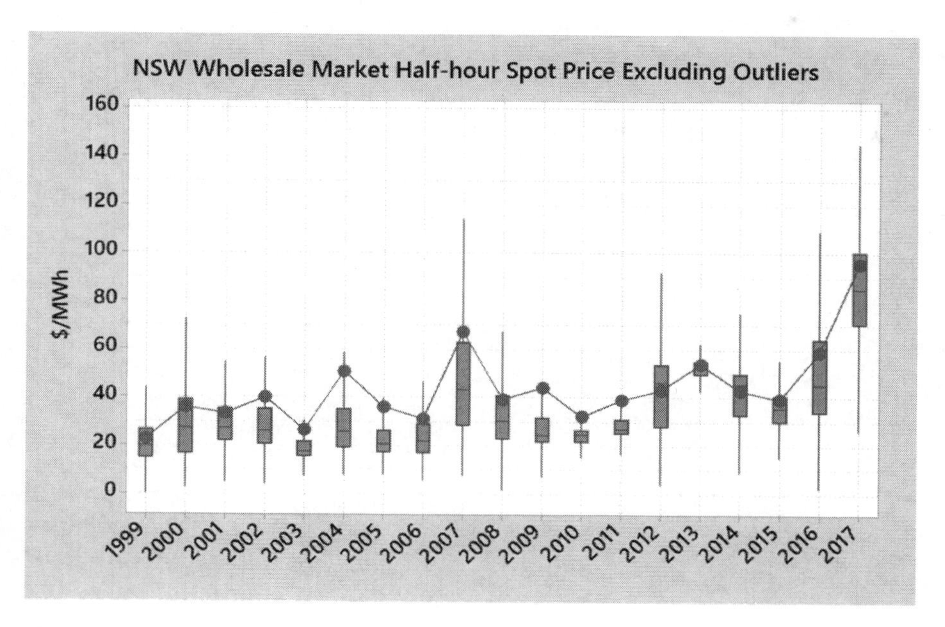

Eighteen years ago the average and median prices were close to $20/MWh or 2¢/kWh. In 2015, the average price had doubled to $40/MWh or 4¢/kWh. This is some of the lowest cost electricity in the world. In 2016, some high price spikes drove the average wholesale spot market price up towards $60/MWh or 6¢/kWh. However, the median price was still down around $45/MWh or 4.5¢/kWh, very similar to the drought period of 2007.

So, roll back to mid 2016 when you are looking to contract for supply from January 2017. Would I accept a fixed retail price of $60/MWh (6¢/kWh) for 2017? Even if I had ability to curtail my load the answer would be, "absolutely, yes", as the pricing would have been a fair offer based on recent pricing and volatility. Would I accept $70/MWh (7¢/kWh)? If I could curtail much of my load to manage the risk of high price spikes, the answer would likely be, "no, thanks" – I'd rather take my chances with the spot market and manage my own risks. If I couldn't curtail, then I would likely still lock that price in.

If I was offered $100/MWh, I would see that as the extreme of all the potential likely future price outcomes, given recent history overlaid with some dark clouds on the horizon with the likely impact of the Hazelwood Power Station in Victoria closing and its impact on NSW prices. I would prefer to opt for the wholesale market pool price and manage the risks myself.

As it has turned out, in 2017 prices escalated across the country due to the Hazelwood closure. The actual price outcome for NSW in 2017 was $95.48/MWh. A wholesale market spot strategy would have provided a modest saving compared with a fixed retail offer of $100/MWh. The other two offers of $60/MWh and $70/MWh would have provided a far better outcome than the wholesale spot pricing strategy.

Victoria

Moving on to Victoria.

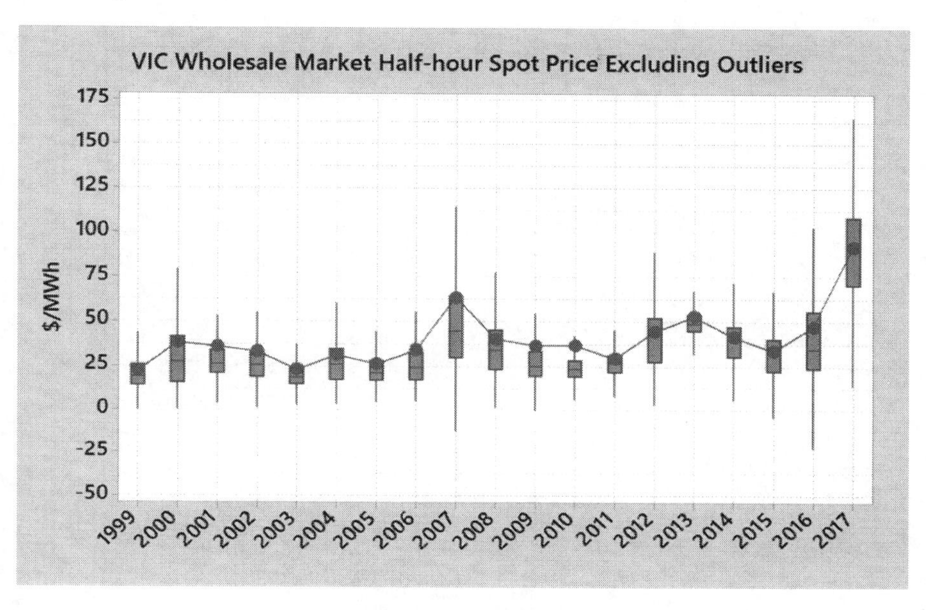

Once again we see the impact of the 2007 drought and the carbon tax period, but look at how steady, predictable and low electricity prices have been in Victoria. Wow. If I was starting an aluminium smelter I would have set it up there. Prior to the carbon price, the spot price median and often the average were below $30/MWh (3¢/kWh). In 2016, the average price was less than $50/MWh and the median less than $40/MWh.

The prices have been so low that it's no wonder the old and carbon inefficient brown-coal-fired station at Hazelwood closed in February 2017. In this case, the future closure of Hazelwood was known in 2016, but not the future actual impact on wholesale prices with certainty. The analysts within the generation and retail businesses would have done plenty of modelling and come up with a range of scenarios, but there would have been a high degree of risk or uncertainty around those forecasts. Who pays for this uncertainty or risk? You guessed it: the customer does. Fixed retail price offers for 2017 factored in a risk premium for this uncertainty. By locking in fixed retail prices when there is uncertainty, you are locking in high prices.

As it turns out, our worst fears were actually realised and the average wholesale market price for Victoria increased to $92.22/MWh for 2017.

In terms of whether 2018–19 fixed retail price offers are fair value or not, it's very difficult to judge based on late 2017 price outcomes. It's difficult to determine how the impact of the closure of the Hazelwood station will play out in the spot market prices in the medium to longer term. The best indication is to look at the current futures prices on the ASX Energy site. At the time of writing in late 2017, futures for 2018 remain above $100/MWh and even $142/MWh in Quarter 1.

So, a retail offer of around $100/MWh in Victoria reflects the current view of 2018 pricing. Anything above $100/MWh is "out of the money" and I would advocate a wholesale market strategy.

South Australia

Next we take a look at South Australia. SA has been prominent in media articles lately due to the closure of the Port Augusta Power Station in mid 2016, the state-wide power blackout in September 2016, and the increasing reliance on renewable generation.

So let's put the facts in a historical context.

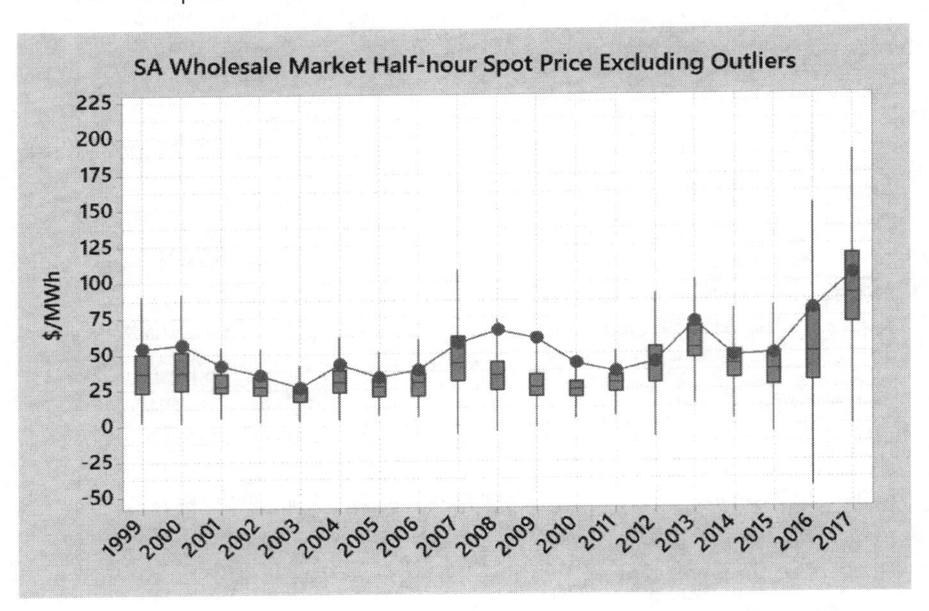

You can see that 2016 was certainly a horror year for spot prices. The closure of the only coal-fired power station at Port Augusta (Northern Power Station or NPS) in May 2016, along with some other factors, sent prices soaring in May through to July, until the Pelican Point combined-cycle gas turbine (CCGT) plant was restarted from its mothballed state and prices dropped back down close to the pre NPS closure level.

The closure of the NPS was not the only factor that drove up prices during this period. The increases in prices were caused by:

- the closure of the coal-fired Port Augusta Northern Power Station in May 2016 (the "headline" issue)
- the "mothballing" of the Pelican Point combined-cycle gas turbine on May 2016 (the "coincidental" factor)
- the de-rating of the Heywood Interconnector with Victoria, which was down due to capacity upgrade work (a "contributing factor")
- a large increase in the short-term trading market price for gas due to LNG export demand (a "structural shift")
- some periods of low wind generation output (a "contributing factor")
- some periods of winter time high demand (a "structural issue").

The next graph shows weekly spot market price distribution, including the impact of the NPS closure and the subsequent restarting of the mothballed Pelican Point CCGT station.

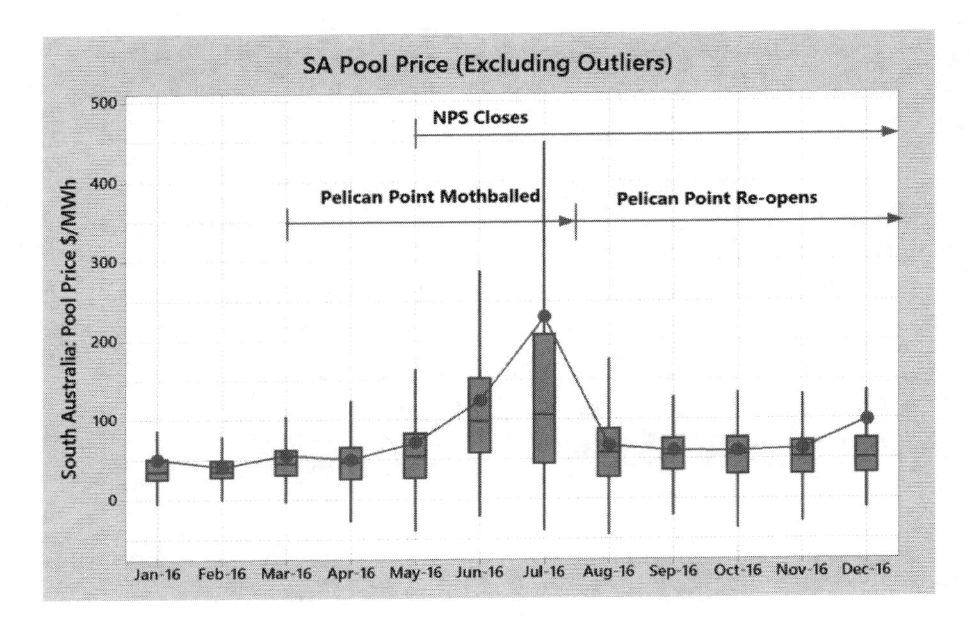

Clearly prices increased dramatically with the closure of NPS but then moderated back to pre NPS closure pricing levels after Pelican Point restarted. However, the media hype through this period, and afterwards, emphasised the increase but largely overlooked the decrease in prices.

The May to July period drove the 365-day average to an unprecedented high level. This price bubble was ultimately reflected in the "low risk" fixed retail prices being offered after this period, but the underlying spot price had already dropped to near previous levels.

So a horror year that saw 7.5 months of no coal-fired generation and unprecedented high gas prices resulted in an average annual spot price of $80.60/MWh (8.06¢/kWh) for 2016. This was the highest annual average since the commencement of the NEM, but – from an end-user point of view – it was not that far above the carbon price period of mid 2012 to mid 2014. It was also not that far above the 2007–08 drought period pricing.

This annual average price, covering both peak and off-peak periods, is approximately the same as the best off-peak prices that were being offered in late 2016 for the 2017 year and well below the annualised flat prices being offered.

When would have been the worst time to go to the market to request fixed retail price offers for 2017 and beyond? Well, after mid May 2016. When do many customers go to the market to request

prices? About three to six months before their contract expiry, right smack bang in the middle of this period. Had SA customers been fully educated regarding what drives fixed retail prices, who pays the risk premium, and alternative options, they may not have accepted the very high fixed-price offers made to them during this period.

By late 2016 fixed retail prices were being offered at or above $140/MWh for 2017.

In all states 2017 has been a bad year for average wholesale electricity prices, including South Australia. However, the average wholesale spot price for 2017 was $105.33/MWh. Quarter 1 was driven by a very high February, when Hazelwood closed. This average price for 2017 was well below the fixed retail offers of $140/MWh being put forward in late 2016. A wholesale market spot price strategy would have yielded significant savings versus the retail offers being made in late 2016.

By looking at long-term historical data we can see that there are sometimes very high price years driven by specific circumstances that do not reoccur every year. These include droughts, long and extreme weather conditions, the shock of power station closures and the subsequent market adjustment, and also specific regional factors. South Australia is, to a large extent, reliant on the interconnections with Victoria to both export and import electricity. A long-term constraint or failure of the interconnector can have large price implications.

So what about 2018?

At the time of writing (late 2017), the futures prices are indicating expectations of around $120/MWh for 2018. However, recent fixed retail offers have been significantly higher than that. A fixed retail offer of $100/MWh would represent good value and should be locked in. Price offers above $120/MWh are "out of the money" and would indicate that a wholesale market strategy would be more effective.

Tasmania

Tasmania is occasionally very reliant on the Basslink interconnector with Victoria. It serves the purposes of both exporting renewable hydroelectric power to Victoria and in providing security of electricity supply to Tasmania in the event of low dam levels. In early 2016 the Basslink failed at a time of relatively low dam levels in Tasmania. Tasmania had been importing much of its electricity needs at the time of the failure, in order to conserve and grow dam levels.

So what happened to market prices after the failure?

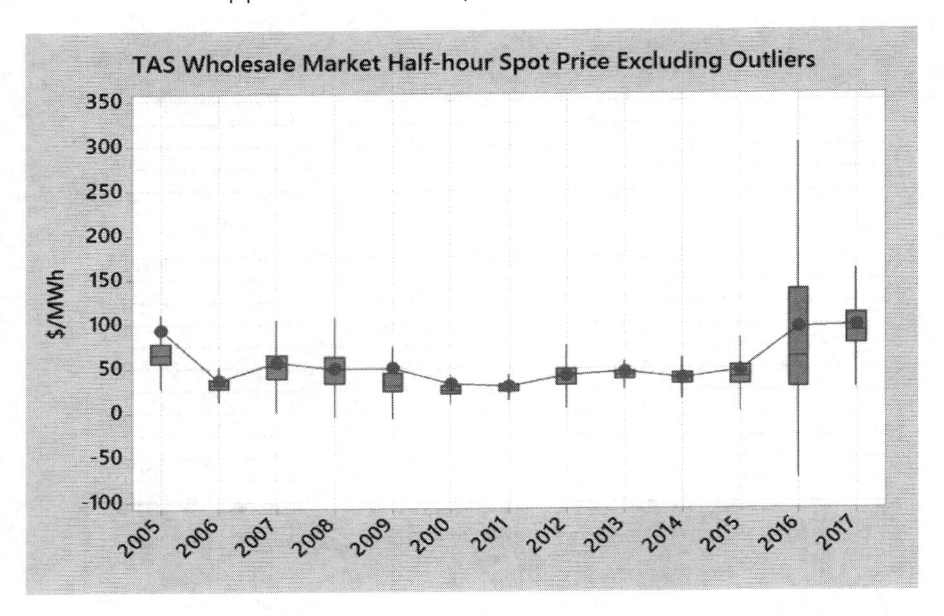

Due to this "once in a decade shock", the annual average price rose to $95.89/MWh (9.6¢/kWh). Was that a bad result for any business that was 100% exposed to the spot price? Yes. Was it a disastrous result for businesses that had been exposed for the last three or more years? No, not a disaster. The spot price over the preceding decade had been around or below $50/MWh.

What is of interest is that while price volatility decreased once Basslink was repaired, the average price in 2017 was still close to that of 2016 at $98.15/MWh. This is due to the Hazelwood Power Station closure in Victoria. I would anticipate that prices would remain around this level in 2018.

Being informed leads to better decisions

The lesson is that a business that is highly exposed to electricity prices needs to understand what is happening in the market and whether fixed retail price offers have built in bloated risk premiums, or whether they reflect fair value. To make this determination, the business needs to understand the drivers of price and the historical context of current pricing.

If you are looking to purchase electricity at wholesale market prices then you should have a good understanding of both the long-term and recent history of average price levels, in order to properly explain the strategy – including opportunities and risks – to the stakeholders in your business.

If you currently purchase electricity at wholesale prices then you should also have a good understanding of history, the recent price performance in the context of that history, and what is driving those prices.

The smart consumer who understands the market knows the reasons for the high prices and can come to an informed view as to whether the factors driving them will start to recede, and if wholesale market prices should be expected to fall, followed by a fall in retail prices. Or they may form a view that the market is entering a new paradigm and they should lock in fixed retail prices now ahead of continuing higher prices.

The very high prices in South Australia in the winter of 2016 after the combination of several factors are a case in point. The informed consumer would have known that the closure date of the NPS being brought forward would have caused a market shock initially, and that it was an abnormal event in the historical context and eventually the market would likely correct itself. In the case of the high 2016 winter prices in South Australia, the owners of the Pelican Point CCGT power station were persuaded by the State Government to turn back on. Immediately prices fell back to below pre-crisis levels. Funny that.

Similarly, the informed consumer would have expected that the closure of the Hazelwood Power Station would cause a price shock in Victoria that would likely spread into New South Wales and South Australia. The worst time to go to market for a fixed retail offer was at the peak of the shock.

If an electricity consumer understands the market, understands historical price distributions and patterns, and understands the drivers of factors affecting the electricity market, they can keep themselves well informed about whether current retail prices reflect value or are priced well above value. They will understand the rewards and risks of taking on a pool price pass-through strategy and the opportunities and costs of including demand response in their strategy.

A consumer that understands the market puts themselves ahead of their competition and takes control of their electricity spend, leading to a better bottom line for the business. A consumer that does not understand the market is a simple price taker and gambles each time they go to the market for fixed retail prices.

CHAPTER 11

Step 4: Understand your electricity bill

Electricity tariffs seem quite complicated and very difficult to understand to most energy consumers. In this chapter we break a typical electricity bill down into its sub-components so that you can understand how a bill is constructed and what the cost levers are, so you can best pull the right levers to reduce costs for your business. We'll also look at the different types of retailers, and how this affects your costs.

The problem with the standard energy procurement process

A typical approach to energy procurement is for the procurement department to engage a consultant to go out to tender or a request for proposal every three years. The consultant will then analyse each of the retailer supply offers and compare the overall cost for each offer over the three-year period. The consultant will include each of the tariff components in their spreadsheet, and then make a recommendation to the client on which offer to accept or negotiate further on. The consultant understands the tariff structure often because they themselves came out of the electricity retail industry.

The business procurement specialist and/or the responsible business unit manager will then meet with the short-listed supplier or suppliers, thump the table a bit, and demonstrate their negotiation prowess by shaving a small amount off the quoted prices and reporting that success back to the business. The successful retailer walks away with smug satisfaction that they didn't have to shave as much off as they thought they would need to. Then the only real interaction the customer and supplier have over the next three years is enjoying a few wines together in the corporate box at the football and cricket, and the odd business lunch.

What the broader business doesn't realise is that the "high pressure" negotiations have shaved a bit off the price involving only about half of the total cost of electricity. The other half of the costs are "fixed" network charges that are regulated and are not analysed anywhere near as much as the non-regulated energy charges.

The energy procurement specialist would be better off skipping the lunch and doing a more detailed analysis of the drivers of the "fixed" network charges, environmental charges and other energy options. It's vital to break down the tariff and understand and analyse each component.

Breaking down the tariff

The tariff is composed of environmental charges, metering charges, retail supply charges, regulated network charges and non-regulated energy charges. A typical network tariff consists of the following elements:

- **energy charges** – the actual electrons that are supplied as power adjusted for transmission and distribution line losses:
  - peak – energy that is supplied during defined peak periods
  - off-peak – energy that is supplied during off-peak periods
  - shoulder – energy that is supplied during shoulder periods (some states)
- **retail supply charges:**
  - daily – a charge that the retailer applies on a daily basis to cover their costs

- **network charges** (a combination of transmission charges and distribution charges):
  - demand – the maximum amount of energy that is used in any half-hour
  - usage – an energy usage rate that the networks apply that varies with usage
  - supply – a fixed daily charge
- **market charges** – pass-through charges from the Australian Energy Market Operator based on energy usage
- **environmental charges:**
  - LRET – large-scale renewable energy target charges based on usage
  - SRES – small-scale renewable energy scheme charges based on usage
  - state environmental charges – state-based renewable energy schemes if applicable
- **metering charges** – a charge for the retailer to engage a metering agent to measure usage (usually a fixed monthly charge).

Energy charges are the most competitive component of the tariff, but there are still some other charges that can create a significant difference between seemingly similar offers.

Breaking down the retailers

Understanding the different types of retailers will also help you to understand pricing and get the best deal for your business. There are generally three types of retailers, but in essence they carry out a similar process:

- The most dominant retailer is a "gentailer". The gentailer is both a generator of electricity and a retailer, and most often has a portfolio of large coal-fired, gas and wind generation capacity. This retailer has an advantage in that they have an upstream arm generating electricity and so can purchase electricity from themselves.

- The second type of retailer is one that does not have any generation arm and purchases their upstream electricity from the market.
- The third type of retailer is one that has a small amount of generating capacity – often gas fired, wind and/or hydro – but has a retail market book much larger than their generation capacity.

One would expect that a customer would get a much sharper offer from a gentailer than the other two types of retailer, but this is often not the case. The reason for this is the way in which the market works behind the scenes.

While a gentailer has both a generation arm and a retail arm, these arms often don't clasp hands together. The two arms act as separate business units and are often headed up by executives who have different business objectives. The generation business unit is motivated to maximise the profit from their generation portfolio. This means selling electricity contracts at the highest possible price, even if that means leaving a large part of their generation fleet idle. The higher the business unit's annual profit, the larger the executives' bonuses will be, and we are talking about bonuses of many hundreds of thousands of dollars or more. The generation arm will sell large blocks of energy to any retail business or intermediary that will buy them at or above their target prices.

The related retail arm is motivated to maximise market share and maximise revenue. It buys blocks of energy from the generation arm on usually a similar basis to the electricity intermediaries. This is a natural strategy, as it's common in many businesses with vertically integrated business units to keep most of the profit as high up the supply line as possible. The further down the line margins are passed, the more the downstream business units "give away" value to the customer in trying to win more market share by competing on price.

The other retailers need to go to the market to buy their blocks of energy to service their customers. Sometimes this is done by purchasing blocks of energy from the competing gentailers or through intermediaries such as banks who purchase from the gentailer.

Factors that affect pricing

Each of the retailers, whether they have an upstream generation arm or not, has access to similarly priced blocks of electricity that are based on the underlying market price. That is why when you pay for a consultant to run an electricity tender for you, many of the retailers come back with quite similar prices despite their different size and generation capacity. There are however differences in the prices which are influenced by a range of factors that the buyer needs to both understand and leverage.

Contributing factors from the retailers

Let's take a look at how retailers contribute to energy pricing.

Volume risk

Firstly, the retailer needs to have some understanding of the customer's load profile in order to procure blocks of electricity themselves. They need to know typical usage patterns throughout the day, throughout the week, and any seasonal impacts. A client will not have exactly the same load from year to year and the retailer needs to take this into account.

The retailer is purchasing fixed blocks of energy at a fixed price and the customer is using variable amounts of energy but still at a fixed price. If the customer uses *less* electricity than the retailer has purchased to supply it, the retailer needs to sell that equivalent amount of electricity back to the market at the prevailing market price, which is highly variable. Similarly, if the customer uses *more* electricity than the retailer has purchased for it then the retailer needs to buy additional electricity from the market at the prevailing price to supply the customer.

The retailer is taking on a volume risk to supply the customer with the amount of electricity that it requires at a fixed price. The retailer reduces the risk from individual customers by aggregating them into a portfolio of customers to reduce the overall volume risk. Some customers will use more and some less, thus offsetting each other in the portfolio. But the retailer cannot completely eliminate this volume risk through a portfolio approach – a residual risk will remain. The retailer quantifies this risk in financial terms and then applies a risk premium to customers in that portfolio to offset exposure to market prices.

When a retailer looks to construct an electricity supply offer to a customer it will analyse the customer's load profile to calculate how that load will impact the portfolio risk. The risk premium charged to that customer will be determined by its impact on the portfolio risk. Very large customers may have their risk analysed outside of a portfolio as the retailer may be buying blocks of electricity specifically for that customer, and so the risk premium will likely be much larger.

The more the retailer understands the customer load profile and the more certainty in the future load profile, the lower the risk and consequently the lower the risk premium. Retailers will typically look at 12 months of load data and make their own determination based on that historical data.

Load profile risk

Not only does the retailer make an assumption about your likely electricity consumption each year, they also make assumptions as to when you will be using that electricity or what your load profile will be. The retailer will purchase contracts for peak, off-peak, and – if applicable – shoulder period electricity. Similar to volume risk, if you consume more or less electricity in the peak period and an offsetting less or more in the off-peak period than what was originally assumed then the retailer will have a surplus or shortfall of electricity and be exposed to the wholesale market price to balance those surpluses and shortfalls.

Credit risk

Retailers are also exposed to credit risk that can be substantial. There are many examples of large and small customers going into administration and then liquidation, owing large amounts of money to their suppliers. The suppliers are at the end of a long line of creditors with a claim on the business. Across the entire portfolio of customers this exposure to credit risk can be calculated and then charged back to the portfolio of customers. Large customers may be required to have a superior credit rating or lodge bank guarantees to reduce their specific risk, but often those large customers will push back and point to their substantial balance sheet in their annual report to argue against the need for lodging security. If they are successful, their credit risk remains for the retailer. The retailer ultimately protects themselves from a credit default by building credit risk into all of the customer price offers.

So your business is paying an insurance premium to protect the retailer from another customer going out of business.

Cost recovery

Retailers are quite justified in recovering their costs when supplying customers. Those costs will vary with the size of the business. They could include the costs of:

- owning or leasing a tall building in a CBD location with their name emblazoned across the building
- floors full of zombie employees in cubicles
- hugely expensive software to manage the large number of customers
- corporate entertainment
- sponsorship of large sporting events or teams
- overseas study trips for key executives
- wining and dining customers, potential customers and business associates.

Or the costs could be as small as a couple of employees housed in a nondescript office in a light industrial area managing a small number customers using a spreadsheet. I have done a 15-year electricity supply contract with a division of a large corporation and that division was two employees, a spreadsheet and a small office.

Whether large scale or small scale, the costs come down to total costs divided by electricity sold. Different retailers will have quite different costs that they will pass on, which will of course affect the prices quoted to your business.

Profit margins

The retailers are also justified in making a margin over their costs. Along with retail costs, this is an area where the retailers are more competitive with each other. The margin that a retailer builds in depends on some quite simple factors that often have a human element associated with them. These include:

1 how large the load is
2 how keen the retailer is on winning that specific customer

3 how sales bonuses and key performance metrics are calculated within the retailer

4 whether the sales representative has a personal like or dislike of the customer or its consultant

5 the history of negotiations between the customer and the retailer

6 whether the customer is going to the market for pricing or simply renewing supply with the incumbent retailer.

* * *

So, energy charges can be represented as:

> **Energy charges = Energy cost + Volume risk premium + Load profile risk premium + Credit risk premium + Retail costs + Retail margin**

Retail supply charges vary between suppliers. They are notional costs of supplying electricity to a specific customer that are often just rolled in with the energy charge, or are sometimes separated as a fixed daily charge. For this reason it's important to not just compare energy charges but also to understand retail supply charges and include these in your analysis.

Contributing factors from the networks

The electricity network is the poles, wires and other electrical equipment that transmits power (transmission) from the generator to the local network, which then delivers power to end users such as households, businesses and industry.

Network charges are the most interesting and complex component of the total cost of the supply of electricity. Network charges are regulated by the Australian Energy Regulator (AER) through a very complex process – it would take a 1,000-page book to cover the basics. So, for the purposes of this book we will just take a brief overview of the basics.

Electricity networks are regulated monopolies

Historically the networks were owned by individual state authorities, which also owned the generation assets and supplied the retail

services. As the states started to privatise the electricity supply industry, many networks were sold off to private enterprises. Those that remained in state hands were turned into state-owned corporations that were required to act commercially. As electricity supply networks are by their nature monopolies, there are very strict laws and regulations in place to protect consumers and stop network owners from pushing prices up.

To make the networks attractive to potential investors, they had to be guaranteed to be profitable and make a return on their assets. This is done by approving the amount of revenue that the network owner can collect from all of the electricity users in its network over a fixed period.

The regulated asset base is determined by the original asset value when the network was privatised, or corporatised, plus any growth in asset value, plus any additional new assets from approved investment.

The level of revenue is determined by the approved regulated asset base value, the required capital spending program to ensure a reliable network, the forecast operating and maintenance costs, and an approved rate of return on the asset base.

As the network owner is guaranteed a return, the business is considered low risk and, in theory, the low risk should have a low return relative to the cost of capital. The AER determines a weighted average cost of capital (WACC) to apply to the regulated asset base. This determines how much revenue can be collected from customers. This is known as the revenue cap.

The network providers submit tariffs for approval that are meant to reflect the cost of supply for different customer classes, with the total revenue not exceeding the revenue cap.

Network providers are also required to achieve certain reliability standards; in simple terms, that means ensuring electricity demand is always met even at periods of high demand. In order to meet the reliability standards the network provider is required to invest in new equipment and operate and maintain existing equipment. So the network owner is incentivised to justify capital expenditure in order to meet reliability standards, and the more they invest, the more they can collect in revenue and the larger their profits.

I'll repeat that. The more they state they need to spend to reliably keep the lights on, the higher the profit that they are allowed to make. That's where the term "gold-plated networks" came from.

The umpire, the AER, makes a determination as to whether the claimed required capital expenditure and operating and maintenance costs are excessive or not, and makes a draft decision on this basis. Invariably the network provider is not happy with the decision made by some of the brightest energy market technicians in the land, and so appeals to the Federal Court through the Limited Merits Review process. The Federal Court often upholds those appeals, forcing the AER to approve the proposed revenue cap and tariffs.

Network charges

In simple terms, network charges consist of a transmission charge and a distribution charge.

As shown in the following image, transmission charges are the cost of transporting electricity from generators to the local network. Distribution charges are the cost of transporting electricity from the transmission network to the end user connection point.

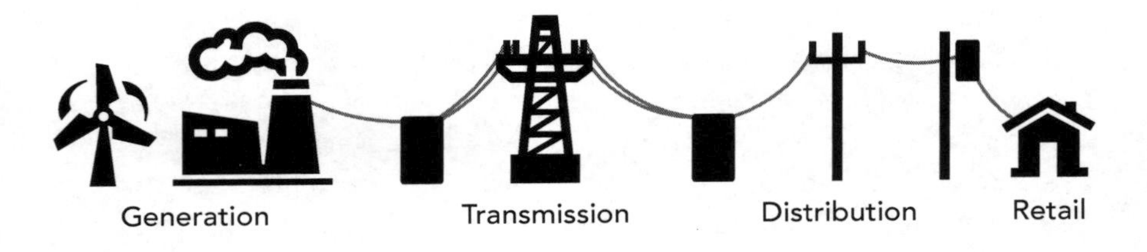

Network charges vary considerably between network providers, however the common features are:

- **A usage charge:** a charge based on how much electricity is used, sometimes based on differential rates between peak and off-peak periods.
- **A demand charge:** a charge based on the maximum demand used in any half-hour period. Demand is usually measured in kVA rather kW, and the charge can be based on the maximum half-hour demand for the month, for the previous month, for the last two months, a 12-month rolling average, or a historical maximum "agreed" demand, depending on the state or network region.
- **A supply charge:** a fixed charge each month.

The ability to reduce network charges varies between networks, however the main lever is the demand charge, with some ability to make savings in some regions by shifting load to off-peak periods. So, depending on the network, substantial savings can be made by:

- reducing and managing your maximum demand during peak periods (see chapter 9, Step 2: Analyse your operations)
- shifting consumption to lower cost periods
- increasing efficiency and reducing consumption.

The "fixed" charges are not actually fixed at all.

The effects of renewable energy

A behind-the-meter solar generation system is going to help reduce network charges as well. This will definitely be through reducing consumption, but it may also help in reducing the maximum demand. The agile consumer may even look to curtail load if solar generation is low at a time when plant consumption is high, to minimise demand.

Environmental charges seem to be the most controversial component of electricity bills. Environmental charges are increasing each year as the country aims for the "20%" renewables target by 2020. The renewable energy target was modified in 2015 to achieving 33,000 GWh of large-scale renewable energy generation by 2020. In effect, this means that by 2020 the amount of renewable electricity generation as a proportion of total generation will be approximately 23.5%.

In order to achieve this national target, the government determines how much electricity is required to be generated by renewable generation each year. In 2017, the large-scale renewable energy target is 14.22%. This means that 14.22% of the electricity that your business uses is required to be backed by large-scale generation certificates (LGCs).

The Small-scale Renewable Energy Scheme operates a little differently – it was designed to encourage households and small businesses to invest in renewable energy and achieve a financial return. Small-scale (<100 kW solar and <10 kW wind) schemes generate small-scale technology certificates (STCs). The scheme generates all of the STCs for a deemed period upfront in the first year of installation. The deeming period prior to 2017 was 15 years, but from 2017 this period reduces by one year, every year, until the planned end of the scheme in 2030. So, the longer you wait to invest in a small-scale scheme, the fewer STCs you will receive.

The annual STC target is based on matching forecast new small-scale generation schemes installed. In 2017, the target was determined to be 7.01%. This means that 7.01% of the electricity that your business uses is required to be backed by STCs.

The retailer is obligated to purchase renewable energy certificates, both LGCs and STCs, to meet their obligations. They then pass this cost through to the end user. There are differences in the pricing offers for environmental certificates between retailers, particularly with LGCs. Some retailers are owned by companies that have their own large-scale renewable generation projects.

There is a lot of media commentary about how renewable energy schemes should be scrapped as they are pushing overall electricity costs up. That may be the case if you simply accept the vanilla retail approach to renewable energy certificates. But rather than just taking a side in the debate, you can take control of your environmental costs through a range of options. You could:

- generate your own renewable electricity to meet your obligations
- purchase the certificates yourself from the market or a third party at a price lower than the retailer is offering
- enter into a long-term arrangement with a renewable energy project to provide you with the certificates, and possibly electricity, at a substantially lower price.

Many businesses have large roof space or large land holdings and so have the ability to install their own solar or wind generation to save costs and generate their own certificates. Or if your business would prefer to allocate capital investment elsewhere, there are alternatives where you could contract for a long-term renewable electricity project for the supply of both electricity and LGCs.

* * *

Step 4: Understand your electricity bill

If you understand the different components that make up your electricity bill you can understand the options available and identify which levers you can pull to reduce your overall electricity costs. This is an important step in your energy strategy.

CHAPTER 12

Step 5: Reduce your network costs

Network costs are often thought of as fixed costs, and most people believe that you can't really do anything to reduce them. In this chapter we will delve into network costs, and you will see that there is actually much you can do to reduce these perceived fixed costs. We will combine Step 2 (Analyse your operations) with Step 4 (Understand your electricity bill) and find costs savings at the intersection of those two topics.

Network cost drivers

As we saw in exploring the components of the electricity bill, network costs are driven by:

- energy use
- maximum demand in kVA (or MVA)
- timing of when the electricity is used.

We can therefore reduce electricity network charges by reducing energy consumption, reducing maximum demand or shifting load from network peak periods to network off-peak and shoulder periods. There is also the option, in some jurisdictions, to change to a more favourable network tariff to suit your operation.

Each network in each jurisdiction has a different method for calculating these charges, and different rates. In South Australia the maximum demand charge (agreed demand) and the additional demand (off-peak maximum demand) are set at the highest half-hour historical demand. The maximum demand is set at the highest historical half-hour demand, in kVA, during the summer period from November to March, between 4:00 pm and 9:00 pm. One half-hour spike some time in the distant past could be setting your demand charges, and that charge may be well above what your facility actually requires.

The additional demand is the highest demand, in kVA, that is set in any half-hour period outside of the summer peak period less the agreed demand level. This charge rate is about 50% of the agreed demand rate.

There is also another option in South Australia to select an actual demand tariff. With this tariff you are only charged the actual maximum demand for the month – but there is a catch. There is both a maximum shoulder period demand – which is between the hours of midday and 4:00 pm – and a peak period demand – which is between the hours of 4:00 pm and 9:00 pm in the summer months from November to March. If you run past 4:00 pm you will incur a much higher demand charge rate than you would have been exposed to under the agreed demand tariff.

An analysis of your actual load profile patterns can be carried out to determine which tariff you would be better off using. For example, if your operations rarely, if ever, run past 4:00 pm then you may well be better off under an actual demand tariff. If you frequently run past 4:00 pm you will likely be better off under an agreed demand tariff.

In other states and network regions, the demand charge can be based on an agreed maximum demand, the actual month maximum demand, a two-month rolling maximum demand or a 12-month rolling maximum demand.

The largest determinant of network charges is the maximum half-hour electricity peak period demand of the business, in kVA. The kVA is determined by the real power, kW, that your business is using and the power quality, or power factor, of your site.

To reduce network charges we can use the load profile analysis to pinpoint drivers of the highest demand and look to see if we can reduce them by improved load scheduling to spread out demand, changing equipment to more energy-efficient equipment to reduce

load, identifying and eliminating anomalies, and by improving the site power factor. We saw examples of these in Step 2 when we introduced the concept of power factor.

Power factor and network costs

So what is "power factor" and how does it impact your monthly network demand charges?

When a motor draws power to rotate and drive equipment it uses "real power" to do the work. However, due to the current waveforms and the voltage wave forms not being in phase with each other we get what is called reactive power that does not do any work. What the network would see for this one load is called "apparent power". When we add up the power-using equipment for the whole site we end up with an aggregate site load in real power and apparent power. The ratio of real power to apparent power is called the power factor. It can be represented in the triangle below.

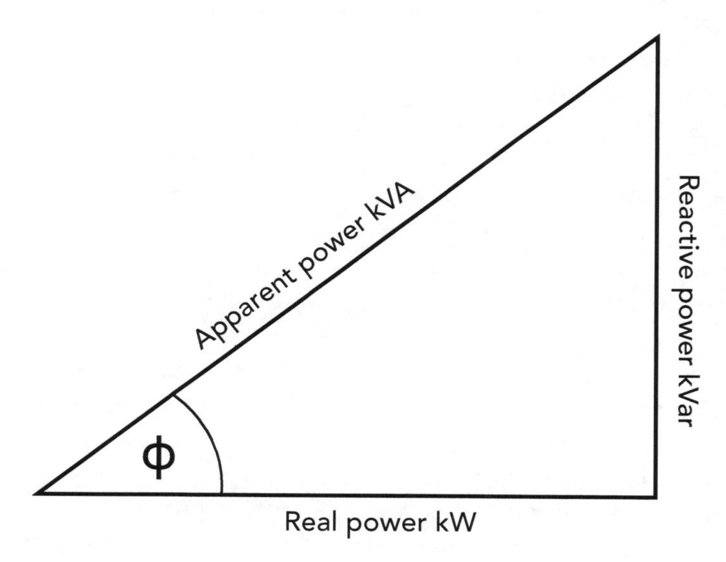

If the power factor is close to 1.0 then the demand that the network sees is very close to the actual load being used. When it is less than 1.0 then the demand seen by the network is larger than the actual load being used to do the work. You pay for the demand seen by the network (kVA) and not the site load or real power (kW).

Some sites have a very poor power factor of around 0.70 or less. This means that these sites are paying 40% more for their network demand charges than is required.

Let's take a real example of a site with a real load of 1,400 kW and power factor of 0.70. Its measured network demand is 2,000 kVA and it pays $0.3189/kVA/day for demand up to 1,000 kVA and $0.2627/kVA/day for demand above 1,000 kVA. This equates to $212,284 per year.

If it installs power factor correction equipment and increases its power factor to 0.98 its new demand would be 1,400/0.98, or 1,429 kVA. It's annual charges would now be be:

$$\$0.3189 \times 1{,}000 \times 365 + \$0.2627 \times 429 \times 365 = \$157{,}533$$

This represents a savings of $54,750 per year.

Of course it costs money to design, buy and install power factor correction equipment (capacitor banks), but experience has demonstrated that the payback period for this modest investment is usually one to two years.

The importance of monitoring power factor

Some sites install power factor correction, which is simply a bank of capacitors, and then just forget them other than, perhaps, some annual maintenance inspections. Capacitor banks can and do fail, particularly the older ones, and the site power factor and maximum demand must be monitored and controlled at all times. It only takes one half-hour period for your site demand charges to be reset to the higher level. If possible, the process control system should be programmed to shed load if any of the capacitors fail and demand increases.

In another example, at a site that I once managed, we'd had a long-term maximum real load of 5,500 kW but poor power factor of less than 0.80, which gave us a maximum demand of around 7,000 kVA. We installed power factor correction equipment in around 2008 and were then able to demonstrate to the network provider that we had reduced their maximum demand in kVA. The network provider agreed to lower the agreed demand to 5,600 kVA based on our demonstrated performance. This represented an annual saving of around $168,000.

Ten years later as an energy consultant, this company engaged me to come back and review their site load profile. What we discovered was that there were a small number of large demand spikes over a 48-month period. When we drilled down into the data we identified that the cause was a sudden decrease in power factor.

The straight line relationship between real power (kW) and apparent power (kVA) shown in the chart below had incidents where the relationship had broken down. The impact on the increase in apparent power is very "apparent".

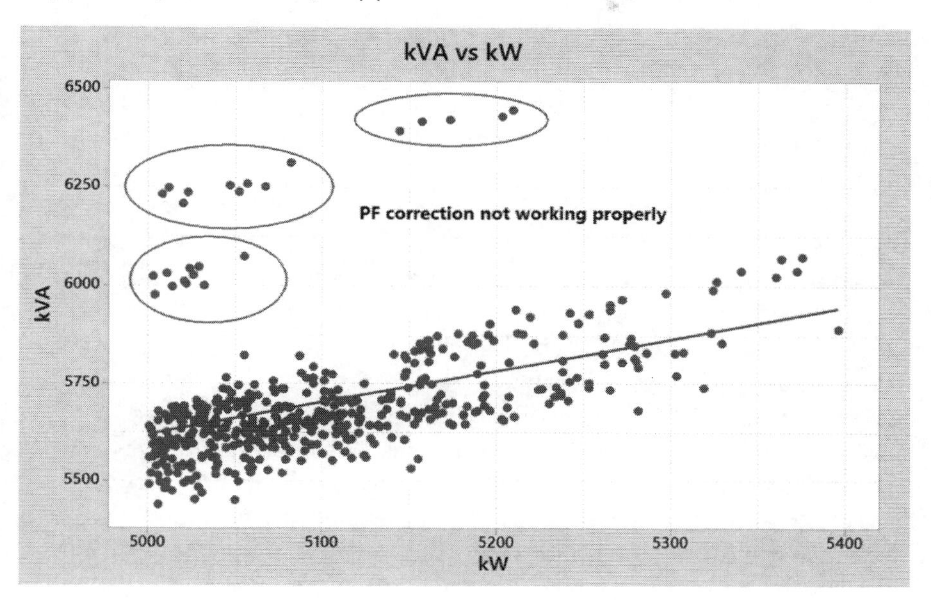

During a workshop to review the results of the analysis the site electrical engineer confirmed that they did have problems with the capacitors from time to time and they did automatically shut down occasionally. What the engineer did not realise though was the impact on site demand charges. As soon as the network supplier measured the increase in demand, the monthly agreed demand level was reset immediately to the higher level and charges increased accordingly. No-one noticed. The site demand charges were increased by $200,000 annually and the operations manager approved the monthly invoice without recognising an increase in the bill and without recognising that anything unusual had occurred. The accounts payable clerk paid the bill without understanding it properly, and the electrical engineer and electricians had no idea of the cost impact of the failure.

These increases can sneak in over time without anyone being aware.

Special causes of changes in network costs

In Step 2 we saw some examples of repeated load profile patterns that were driving up demand costs. Often, there are one-off special causes that also drive up network demand costs. Special causes occur when data falls outside the normal variation of a process. Special causes occur regularly in electricity data and should be investigated and eliminated. Even better, they should be detected as they occur and preventative actions taken immediately to reduce their impact.

The failure of power factor correction capacitors are special causes, but there are many more. One quarry site that I was asked to analyse had closed down part of their operations 12 months earlier and they had requested a reduction in the agreed demand from 600 kVA to 120 kVA. The network provider examined their demand data and reduced their agreed demand, but only to 261 kVA because there had been one half-hour period where demand spiked to 261 kVA. The business was unaware that the agreed demand was not reduced to the full amount that they had requested. The additional annual cost due the full reduction not being made was $18,000 per year.

The reason for the full request not being agreed to is very clear in the chart below. Three half-hour periods spiked above the normal demand level. Six months later there was a similar spike but to a lower level.

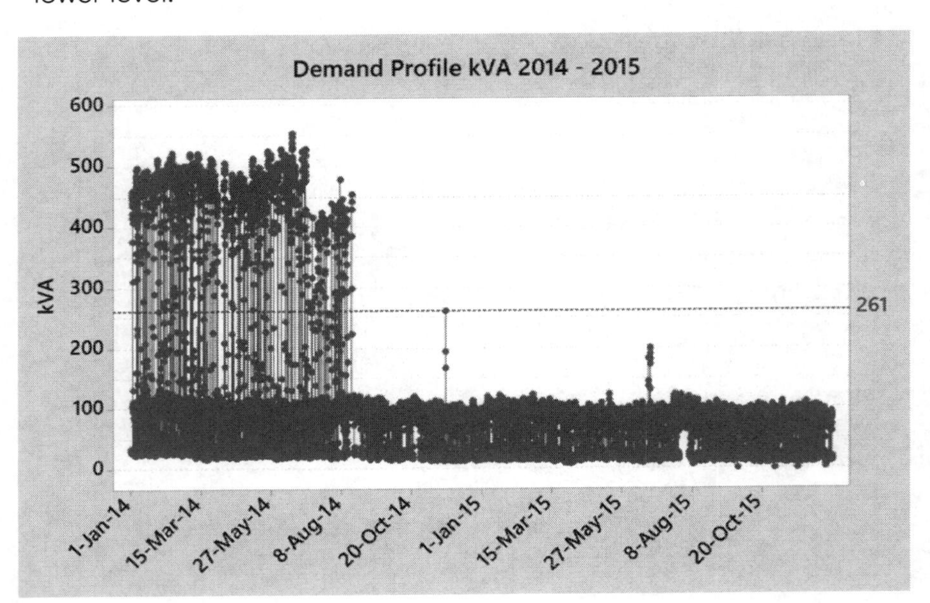

On drilling down into the data on the day of the high load we discovered that someone had decided to start up the part of the crushing circuit that had been closed down 12 months earlier to run out the material that had been left on it. They decided to do this in the middle of the afternoon, during peak period.

There are several lessons to be learned from this example. One is that it is important to monitor load continuously. Every site should have visibility on the load they are drawing at all times to detect if there is an abnormal state. Another is the importance of educating employees about electricity cost drivers. No-one even thought of the implications of starting the crushing circuit up for a short period to run material out. Thirdly, someone in the business needs to be accountable for checking each bill in detail and tracking initiatives to reduce charges.

Common causes of changes in network costs

Once you have identified the special causes that have resulted in short-term spikes in demand and eliminated them, you can start to turn to the "common causes" of high demand. "Common cause" variation is due to the natural variation in the system, which can be quite wide or very tight. This variation is like the natural rhythm of the equipment. There are periods it runs at a higher demand, periods that it runs at a lower demand, and then everything in between. Nothing special has happened and the equipment is being run as per normal.

Reducing demand by looking at the common causes of variation and then either eliminating some of them, substituting them or controlling them can help reduce demand charges substantially.

One of the easiest ways to reduce demand is to audit the equipment you have in the business and what the power draw is for each piece of equipment (measured through sub-board or individual equipment monitoring), and then analyse whether you can reduce that equipment with a more modern, lower power draw and more efficient equipment. This load study can be a part of the scope of the energy-efficiency audit conducted in Step 7 (chapter 14, Reduce your electricity consumption and improve efficiency).

An example of this is when a client who operated a large quarry operation conducted a review of its plant air requirements. We identified that it had two old and oversized air compressors that had been handed down, second-hand, from another site within the company to save on capital costs. A 30 kW compressor was replaced by a 15 kW compressor and a 75 kW compressor was replaced by a 33 kW compressor. This was a combined reduction of 57 kW in load. The site had a poor power factor and so the demand reduction was 70 kVA. Aside from the electricity efficiency savings, the annual network demand savings were around $10,000. With the addition of maintenance savings the payback on the capital of these compressors was less than one year.

This is another good example of where operating costs were unwittingly increased over many years by trying to save on initial capital costs. Analysing the size suitability of current equipment, particularly air compressors, can identify opportunities to reduce network costs.

Identifying opportunities

We will cover electricity efficiency savings in Step 7, but it is important to understand that much of the lifecycle costs of electric motors are in electricity costs and that the capital cost is just a small component of the overall lifetime cost. Identifying inefficient or oversized electric motors can lead to substantial decreases in both electricity consumption and network demand.

Many of these opportunities have very short payback periods, often less than one year. However, some of these types of capital investments may have longer paybacks of up to four to five years. This is where we refer back to the business strategy to determine whether these investments should be made. If a business has identified rising electricity costs as a major threat to their long-term profitability and their stated objective is to reduce electricity costs by 30% or 50% then these types of investments are completely in line with the strategy. The projects should be evaluated on a net present value or internal rate of return basis, not just a simple payback. If electricity costs are not identified as a strategic issue then projects that have a payback period of more than two years will likely be passed over by other business investments. But if that's your approach then you would be unlikely to be reading this book anyway.

We looked at a list of potential strategy objectives earlier in the book:

1 Reduce overall energy consumption by 30% with smart investment in energy-efficient technologies by June 2018.

2 Reduce overall energy consumption by 10% by being more efficient without capital investment by June 2018.

3 Reduce energy costs by 50% by June 2018.

4 Secure fixed-price energy supplies for the next three years at the lowest possible price by June 2018.

If we have a project that reduces demand and consumption that has a payback of four to five years then obviously strategies 1 and 3 would be served by investing in these projects but strategies 2 and 4 would not. It all comes back to aligning your business decisions with your business strategy.

Monitoring your maximum demand

Different networks have different mechanisms for reducing the demand charges. Some are automatic as you pay for your actual demand for the billing month, the billing month prior, the maximum over the last two months or the 12-month rolling period.

In South Australia you must apply to the distribution network provider via your retailer to reduce your maximum demand and/or your anytime demand. With this application you need to state the reason why you are reducing your agreed demand and they will analyse your recent data to ascertain whether they can reduce your demand to that level. You must have demonstrated that you have reduced your demand. If it is demonstrated that you have reduced your demand then they will likely agree to reduce your agreed demand level.

There are a couple of catches though. If you exceed your new agreed demand (or additional demand) in the 12 months after the change then your demand will be reset to the higher figure and you will be retrospectively charged the difference from the date of your change. The second catch is that if you wish to invest in new plant in the future and if the network supplying you is at capacity then you will need to pay for any network equipment required, such as a new transformer, to support your increase in demand, even if it only takes you back to the previous demand level.

If you are in a location where the network is close to capacity and where there are other major loads, or future loads, and you are looking to potentially expand in the next few years, it may not be prudent to reduce your agreed demand as you may have to pay more to get it back again. In a practical sense though this is unlikely to be an issue for most businesses as there is often sufficient network capacity and small increases are not likely to be an issue, whereas large increases would have required that same investment anyway.

This next step is critical. You must continually monitor your maximum demand and ensure that special causes do not drive demand back up and negate the cost reductions that you have achieved. If your operations have a central process control system then you can set it to automatically shed load if a maximum set point load is being approached.

The example of the power factor correction equipment failing highlights the point. If your capacitor banks fail for some reason, your load stays the same and the plant continues to run without anyone noticing anything, at the end of the month you will see an increased demand charge that will carry through to subsequent months depending on the network that you are in. Even then, nobody may notice as it may be obscured in the other cost noise.

The other advantage of continual monitoring is that it provides a measure against which you can assess your business and create internal motivation to continually improve against it. A challenge can be set to reduce maximum demand each month. It is amazing what initiatives employees can come up with when they are motivated towards taking on the challenge.

* * *

Network costs are not fixed. The demand component can be significantly decreased by identifying the common cause patterns and the special cause outliers. Once these are isolated, initiatives can be implemented to reduce the maximum demand or shift it to lower cost periods. Power factor correction will decrease demand charges and often have payback periods of less than two years.

Monitoring and control of maximum demand is critical to ensure that demand charges are minimised and any savings locked in.

CHAPTER 13

Step 6: Reduce your energy unit costs

Retailers bundle up electricity products into a vanilla product, and many businesses are unaware that there are other options besides this product. This chapter looks at an alternative to buying electricity at risk-premium-laden fixed prices, and shows how significant savings can be made by buying electricity at wholesale market prices. It will also highlight the risks of this approach, and the methods of managing that risk.

The standard retail contract

Most businesses in Australia purchase electricity as a bundled retail product from a licensed retailer. The retailer takes care of the administration of the customer's supply service, so that the customer need only turn on the switch and electrons start flowing, and a bill turns up each month.

The retailer is doing a lot of work in the background that the customer never sees, such as:

- purchasing electricity to supply to the customer
- purchasing the environmental certificates required to meet the customer's obligations

- organising a meter agent to monitor, record and report consumption data from the meter
- paying the network provider for transmission and distribution charges
- calculating the bill and sending bills to customers, and ensuring that they pay.

All of this work takes human effort as well as automated processes, and therefore there are associated costs. Retailers are not in the business of providing free services; their owners – government, local shareholders or overseas companies – require the retailer to make a profit, so they charge a profit margin on the provision of the electricity supply service. All of this is fair and reasonable, as it would cost most businesses more to do this work themselves, unless they are extremely large electricity consumers.

The hidden cost

However, retailers provide another service that most businesses don't even realise they are getting, and it's an expensive service. Retailers provide a risk management service bundled in with the supply of the electricity, often at high risk premiums.

There are quite a few risks that the retailer takes away from the customer, including:

- **volume risk:** the risk of exposure to the wholesale market if the customer does not consume the exact amount of electricity assumed in the supply agreement
- **load profile risk:** the risk of exposure to the wholesale market if the customer does not use electricity throughout the day, week, month and year as was assumed in the supply agreement
- **market risk:** the risk of exposure to fluctuations in prices in the electricity market
- **credit default risk:** the risk to the retailer if the customer cannot pay their bills.

To some extent the retailer is smoothing out the risks by aggregating all of the different customers into one or more portfolios, thus reducing the large number of individual risks into

one or more portfolio risks. The profit risk to the retailer can then be stress tested against different scenarios, such as large movements in average market prices and changes in the overall economy that affect demand and credit defaults. The impact of the assumptions made in an individual supply contract can also be tested against the overall portfolio.

So the poor old retailer is bearing the brunt of everybody else's risks. As the retailer is having to wear those risks, it charges a premium to do so, and that premium reflects worst-case scenarios. Instead of the retailer being burdened by the weight of their customers' risks, it is now feeling the weight of potential sacks of gold over its shoulders. If the worst-case scenarios do not manifest then they get to pocket the worst-case scenario premiums that they have charged their customers.

The customers who enter into a straightforward vanilla electricity supply agreement pay the following price for the energy component of their bill:

> Retail energy price = Wholesale energy price + Administration costs + Risk premium (volume, load profile and credit risks) + Retailer margin

There are other options

Many businesses don't even realise that there are other options for purchasing electricity. Negotiating a good price on a standard retail contract might seem like a good approach, but if you understand the intricacies of the electricity market and the power requirements of your business, you can bring huge financial benefits to your business if you get a bit more creative with your electricity purchasing.

Managing your own risk

The risk premium that the retailer builds in assumes that the customer has no ability to manage its own risk. However, there are ways that the customer can manage its own risk through demand side management (DSM) to avoid high prices and achieve very low average prices compared to what is offered by the fixed retail offers.

Load curtailment

The first DSM method is load curtailment. By monitoring market half-hour prices, businesses can reduce their load when price spikes occur and avoid the very high prices that sometimes drag the average monthly price above the median price.

Load shifting

The second DSM method is load shifting. By understanding likely price patterns, such as early morning and late afternoon price increases, businesses can often shift their load – to some extent – to maximise the exposure to the very low prices and minimise their exposure to the higher prices.

Great examples of this are irrigators and water utilities that can schedule pumping at times when prices are historically low, while not running when prices are historically higher. Couple that method with load curtailment to avoid the price spikes and these businesses can achieve average prices well below the fixed retail price offers.

Adopting a pool price pass-through strategy

So, your business carries out an analysis of the electricity spot market, your own load profile and operating constraints, and weighs up the opportunities and risks of adopting a pool price pass-through strategy. But where to from here?

There are two ways you can adopt a pool price pass-through strategy:

- The first is to contract with a retailer to supply you on the basis of pool pass-through prices. With this type of supply arrangement the retailer will charge a management fee to cover their own costs and to make a margin.

- The second way is to become an electricity market participant as an end-user participant. This method is usually only viable for very large users and requires significant management oversight.

Let's consider each of these options.

Wholesale prices through a retailer

There is an alternative to paying what can be very large risk premiums which most businesses don't even know about. Businesses can buy electricity at wholesale prices through a retailer

and keep the risk premium in their own pockets. This leads to a new price calculation for the business:

> Wholesale price = Market pool price + Administration costs + Retailer margin

The key difference is that the market pool price is not fixed; it fluctuates every half hour, and the average cost will vary from month to month. The other obvious difference is that the customer is not paying the risk premium.

Many of the large retailers are not interested in providing a pool price pass-through product to customers. Aside from the fact that they may make less margin as the risk premiums that they usually receive are no longer part of a bundled supply agreement, their billing systems are often not set up for it. It may require someone to manage a single customer with their own spreadsheet outside the normal billing system. They also may consider that the credit risk, or risk of default, is higher. They therefore usually only, begrudgingly, offer this type of facility to very large customers, and usually with a large management fee.

Small and medium-size retailers are often more flexible and willing to offer a pool price pass-through arrangement. The management fee varies considerably between retailers, and it is important to find one that provides the best service and/or price. You can go out to tender or make a request for proposals from retailers for a pool price pass-through supply offer, and then simply compare the product offerings. This will be largely, though not solely, based on price.

Some retailers are becoming innovative in this space and are offering a product that allows the customer to switch between pool price pass-through and fixed retail each month, or to take a combination of both. Some retailers provide price alerts and other information services that add value to their offer.

Going directly to the market

A large user may opt not to pay the retailer any management fee and become a market participant themselves. This is something we did in one of the businesses that I worked for. We had already established ourselves as market participants in the gas short-term trading market, and the leap to doing something very similar for electricity was not very large. However, inevitably there is still a substantial cost in managing the obligations that a market participant has, whether you outsource this function or employ someone and bring it in house. If the cost of becoming a market participant is not too different to the retailer management fee then I would suggest going with the retailer product instead.

* * *

If you contract with a retailer to provide you with a pool price pass-through product then your monthly billing will be very similar to the standard billing you would get with a vanilla contract. The only difference is that instead of peak and off-peak energy usage charges you will be given an energy pool price pass-through charge.

It's important that you have your own internal check of what those charges should be, so that you do not simply accept the retailer charges. The retailer charges should not be a surprise as you should be monitoring and reporting every day on your electricity usage and costs, and the end-of-month bill should be confirmed with your own data.

Our previous analysis has shown that if you adopt a pool price pass-though strategy your average price for the full year is likely to be at or below the best fixed retail price. Historically, in South Australia it is likely to be significantly below, and depending on your load profile it may well be lower in Queensland. In Victoria and NSW the average price is likely to be near the fixed retail price.

Developing a load curtailment schedule

So how do you develop a load curtailment schedule that optimises your electricity cost without impacting your customer service and supply of products?

The Theory of Constraints

This is where we turn to the Theory of Constraints. The Theory of Constraints posits that with any manageable system – for example, a plant – there is one constraint, sometimes more, but rarely many. This means that there are usually many processes that are not constrained and can be turned off, or curtailed, without impacting the efficiency of the total supply chain.

We need to identify the critical asset or assets that are constrained and then determine the opportunity cost of shutting

down the asset, if at all possible. The opportunity cost establishes a break-even point where if that asset were to not operate for a short period of time then the opportunity cost equals the financial benefit of operating.

There are of course other considerations. The customer should see no impact from curtailing load and should be oblivious to the fact that you do curtail as part of your operations management. Also, some types of equipment – particularly equipment that runs at high temperatures – are not designed to be shut down and restarted for short periods of time. This equipment may experience thermal shock, which will have a deleterious impact on equipment integrity or shorten its service life. For example, equipment that operates at elevated temperatures is usually refractory lined, and thermally cycling this equipment will have an impact on refractory life.

The other non-constrained assets are those that are not required to run all of the time. This equipment is often shut down because the buffer stock after its process fills up or its scheduled operating time is less than the constrained asset run time.

Two load curtailment examples

Let's take a look at the example of a cement plant. The following diagram shows the key unit processes.

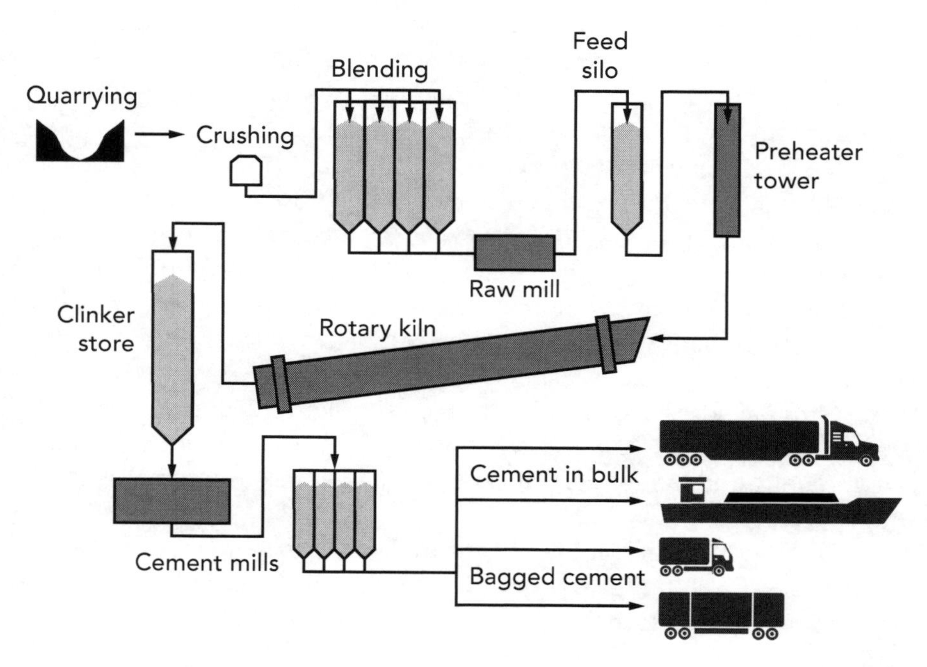

The heart of an integrated cement plant is the clinker rotary kiln. The kiln has internal flame temperatures in the order of 1,500°C and so is refractory lined. Turning the kiln off and on for short periods has a negative impact on the refractory lining and shortens its service life. It is also the constrained asset. It is usually required to run 24 hours per day, seven days per week for more than 340 days a year. The kiln is usually only purposely turned off once a year for two to three weeks for an annual shutdown to replace refractory material and repair or replace other wear items. A curtailment strategy would usually only require the kiln to come off under extreme circumstances.

The processes prior to the kiln involve the mining, crushing and processing of materials. Quite often it's possible to turn off the crusher and raw mill for short periods without having any impact on the kiln.

The cement mill or mills are usually designed with greater capacity than the kiln and greater capacity than the cement product demand. It is therefore very easy to instantaneously turn cement mills off without impacting customer supply, provided cement silo stocks are adequate. There is also a large amount of ancillary equipment that can be closed down with no impact on customer supply.

It's therefore possible much of the time to reduce the electricity demand of a cement mill by 50% for short to medium periods without impacting customer supply or profitability. In some cases, cement not made locally due to high price events can be replaced by cement made at other facilities, or even imported cement.

I've had many clients who initially say there's no way they could contemplate shutting down equipment even for half an hour as it would impact upon costumer supply or service. However, when I have analysed those same plants I have found that they consistently shut down every day for morning and afternoon tea, lunch, meetings, breakdowns, and for full stock. They were often surprised at how often they actually shut down.

Other clients have asked what they should do with their labour when they do shut down. There is always opportunity for clean ups, communication meetings, training, equipment risk assessments, and the list goes on.

On the other end of the spectrum, I've had clients that have said they would shut everything down when price events occur at a particularly high level; for example, $1,000/MWh. This strategy

may be effective if capacity exceeds demand, but it does not allow the fine tuning of cost reduction by curtailing equipment that's not required when prices rise above a lower threshold; for example, $100/MWh.

Next, let's examine the impact of demand side management for a hypothetical plant in South Australia, using actual 2016 prices over the whole year.

In order to produce its target production, the plant aims to consume 8,784 MWh of electricity. However, it has total flexibility in when the plant is run, and can increase and decrease the number of production lines to meet product demand.

The possible load profiles are:

- **Profile 1:** run 24 hours, 7 days a week
- **Profile 2:** 8:00 am to 4:00 pm Monday to Friday, excluding public holidays
- **Profile 3:** 8:00 am to midnight, Monday to Friday, excluding public holidays
- **Profile 4:** 24 hours a day, Monday to Friday, excluding public holidays
- **Profile 5:** 8:00 am to 4:00 pm, 7 days a week
- **Profile 6:** off-peak operation only (nights and weekends)

	No Curtailment				Curtail @ $1,000/MWh			
	Load MW	MWh	Cost $	$/MWh	MWh	Cost $	$/MWh	Count > $1000
Load Profile 1	0.500	8784	$707,899	$80.59	8,736	$574,791	$65.80	96
Load Profile 2	2.186	8784	$888,664	$101.17	8,712	$695,618	$79.85	33
Load Profile 3	1.0903	8784	$921,015	$104.85	8,703	$709,745	$81.55	74
Load Profile 4	0.728	8784	$815,160	$92.80	8,717	$627,686	$72.01	92
Load Profile 5	1.500	8784	$754,082	$85.85	8,735	$621,594	$71.17	33
Load Profile 6	0.834	8784	$530,219	$60.36	8,762	$460,673	$52.57	26

We can see the large difference in electricity costs with different load profiles. Load Profile 3 (8:00 am to midnight, Monday to Friday) was the most expensive and would have cost $921,000. This would have cost $104.85/MWh or 10.5¢/kWh.

Shifting the load to off-peak periods (Profile 6) would have resulted in an almost $400,000 saving, despite the same electricity consumption. This would have cost $60.36/MWh or 6.0¢/kWh.

Let's examine the impact of load curtailment at prices exceeding $1,000/MWh. For simplicity, we will assume that the curtailed load was not made up, and that the full value of the curtailment was achieved (load was curtailed to zero for the full half hour). This would not be the case in reality, but the example serves to show the full "size of the prize".

The savings achieved with load curtailment ranged from $70,000 for Load Profile 6 to $211,000 for Load Profile 3. The savings were, on average, 20% below the non-curtailed load profiles.

Obviously you must consider the impact of curtailment on production. Load Profile 1 would have lost 96 half-hour periods over the year, equivalent to two full days of lost production. Load Profile 6 would have lost 13 hours over the full year.

In reality you would likely set up a curtailment schedule that curtailed some of the load and not all of it, and on some occasions customer demands would prevent you from curtailing and on other occasions the price spike might be missed altogether.

This example serves to illustrate the magnitude of the savings possible through the DSM techniques of load shifting and load curtailment. It also serves to show that even though very high prices might occur, over the full year these are averaged down by the majority of low prices. In fact, during this 2016 period the maximum half-hour price was $13,766/MWh.

Understanding risk

But, *isn't it risky not to use standard retail contracts?*

Going a different way does have risk, but risk does not mean risky. Risk is simply uncertainty. With a pool price strategy the average annual price is not certain and so there is a risk. A retail price is fixed and so there is no price risk. However, there is a very high risk that the fixed retail price will be higher than the average wholesale market price over a full year.

So yes, a spot exposure strategy is more "risky" than a vanilla fixed retail contract, and with taking on that "risk" there is an expectation of a return. History has shown that the worst possible outcome from a pool price exposure strategy has been close to the best retail offer prices, depending on the timing of the retail offer.

In the wholesale electricity market the risk is the distribution of average prices. This risk can be quantified through looking at historical results and by forecasting into the future. The very clever propeller heads in the generation and retail segment do this all the time. The fixed-price retail offers reflect the best known information and projections at the time of the offer.

The trick for businesses is to understand historical price volatility and annual average prices, overlay your informed view of what is happening in the market structure on the coming years, and then compare that with fixed price offers. If your informed view of likely market price distribution outcomes is at or above a fixed price offer then by all means lock in the fixed price offer. If your informed view is that market prices are likely to average below the best fixed price offer then why would you want to lock in high prices?

I think it is risky for businesses to lock in high fixed price contracts without having regard for the likely outcome of variable market price outcomes. The clever businesses realise this and embrace the so-called "risk".

Electricity price risk management strategies

Clever businesses understand risk and so are able to develop and implement risk management strategies to either further reduce their costs or limit downside risk. Electricity price risk can be managed using many different strategies:

- a retail contract fixes the price and gives certainty, but includes a large risk premium
- a wholesale pool price exposure with load curtailment reduces exposure to high price spikes but the final price is not certain
- hedging with financial instruments can provide some protection, such as:
  - exchange-traded (ET) futures, options and caps
  - over-the-counter (OTC) swaps, options and caps.
- physical hedging using self-generated electricity during price peaks; for example, solar, diesel, gas fired
- hybrid retail contracts.

Let's have a look at each of these.

Retail contracts

As we've looked at throughout the book, retail contracts can vary in terms but they essentially fix electricity prices on a flat or peak and off-peak basis for variable volumes. Such retail contracts eliminate price risk for the end user as the end user is paying the retailer a risk premium for the retailer taking on spot risk and volume risk.

As we've seen, the price paid for electricity retail contracts consists of a number of components – these include the price paid for the energy (kWhs) inclusive of the retailer costs and margin, and the insurance policy (risk premium) that protects customers from the volatility of electricity prices. Most consumers do not recognise the magnitude of this insurance premium, or even that they are actually paying the retailer an insurance premium.

Tom Adams, a former executive director of Energy Probe, an industry watchdog in Canada, has a great analogy about paying for fixed retail contracts: "Over a lifetime of driving, you know you'll replace a couple of windshields. But if you insure your windshield, you don't pay for two windshields – you pay for 10."[4]

In individual years a wholesale market price exposure may do as poorly as a fixed retail contract, or worse, but over the course of 10 years you will almost certainly be well ahead of fixed retail prices. This has been my own experience over 16 years since 2001. The drought year of 2007–08 was unexpected and wholesale market price exposure performed worse than fixed-price contracts. And, depending on the timing of going to the market, 2016–17 has performed worse than fixed retail offers due to the bringing forward of the closures of the Northern Power Station and Hazelwood Power Station.

Wholesale market exposure with load curtailment

Customers choose exposure to wholesale (spot) pricing when they realise they will achieve lower electricity costs in the long run by accepting the risks of volatility themselves, and can also provide themselves with their own physical (load curtailment) insurance and financial (hedge) insurance.

4 The Electricity Forum, 2002.

To achieve lower pricing in the long run, the procurement of electricity is no longer a simple procurement process involving competitive tenders but one more complex, requiring an integrated approach between procurement, operations and risk management. I have found that there are often opportunities to curtail a significant portion of total load without impacting on business objectives such as total production or maintaining customer supply or service.

Hedging with financial instruments

Hedging contracts are a means by which participants can negotiate or lock in pricing certainty over time over a fixed volume of electricity. Hedge contracts can be traded on a financial market such as the ASX Energy futures exchange or over the counter (OTC) with financial institutions.

These contracts settle financially at the end of the contract period based on the difference between the contract value and the average price settled in the wholesale market.

Exchange-traded products

Exchange-traded futures and options are traded on the ASX Energy futures exchange. A futures contract is an agreement to buy or sell an underlying asset in the spot market at a predetermined time in the future for a fixed price that is agreed today. Essentially a block of electricity to be used in the future can be purchased or sold in the financial market today at an agreed price. This allows an end user, retailer or generator to lock in future prices today.

The futures contract is a financial instrument, or derivative, based on an underlying physical commodity, in this case electricity.

There are four futures products listed based on base-load and peak-period electricity for calendar quarters for the states of New South Wales, Victoria, South Australia and Queensland. These quarterly futures products can also be traded as a strip of futures representing a calendar year or a financial year. They are:

- monthly base-load futures
- quarterly base-load futures
- quarterly peak-load futures
- quarterly base-load $300 cap futures.

For example, if in December 2017, March 2018 base-load futures are trading at $100/MWh and a factory decides that they would like to lock in that price now, they could buy that futures contract.

Let's say that the factory has an average load of 1 MW. The 1 MW March 2018 futures contract is purchased for $100/MWh in December 2017. This contract has a value of $100 × 31 days × 24 hours × 1 MW = $74,400. However, the purchaser does not pay the full value but rather puts up an initial margin that is usually around 5% of the value (this can change with market volatility). In this case the purchaser needs to put up a margin of 5% × $74,400 = $3,720.

As the futures price is settled each day, the margin account is credited or debited depending on which way the market has moved (in favour of the buyer or the seller of the contract) until the expiry of the contract period, in this case at the end of March.

If the average spot market price for March 2018 settles at $110/MWh then the contract purchaser is paid $110 × 31 × 24 × 1 = $81,840, making a "profit" of $7,440 on the futures contract that they had purchased for $74,400.

The factory paid an additional $10/MWh for their actual electricity consumed more than their $100/MWh target as the average price was $110/MWh. Therefore, this electricity cost them an additional $10 × 31 days × 24 hours × 1 MW = $7,440.

The profit or gain on the futures contract offsets the additional cost of the actual electricity used above their locked-in price of $100/MWh.

If the actual average spot market price settled at $90/MWh for the month, the factory would have paid $7,400 less for their actual electricity consumption versus their locked-in price of $100/MWh, but would have made a loss of $7,400 on their futures contract.

The gain or loss on the futures contract offsets the additional or lesser cost of the physical, or actual, electricity consumed.

The three major risks of such a strategy are:

- *Basis risk:* the hedge is not a perfect hedge as the futures contract is settled on the time-weighted average price whereas the price paid for the physical is a load-weighted average price. If the actual load is higher during high price periods and lower in low price periods the load-weighted average price of the physical would be higher than the settled price of the futures contract.

- *Over-hedging:* in this case the physical was slightly over-hedged by one contract compared with actual consumption and slightly reduced the effective cost. This risk is mitigated by under-hedging the likely forecast electricity consumption.
- *Default risk:* this is the risk that the futures contracts will not be settled. This risk is very low with exchange-traded futures as a market clearing house novates the contracts and eliminates counterparty risk. ASX participant rules in regards to participant financial minimum requirements, initial margin requirements and daily mark-to-market requirements ensure that the contracts will be settled. Over-the-counter contracts have a higher default risk than exchange-traded contracts, however this is also usually very low.

A quarterly base load $300 cap is a contract that is similar to the futures contract we just discussed, but the average price over the quarter is calculated such that the maximum price in any one half-hour interval over the quarter does not exceed $300/MWh. This allows the end user to capture the benefit of lower prices but limits the downside half-hour price to $300/MWh.

There are also options traded on the ASX for electricity contracts. A call option gives the holder the right, but not the obligation, to buy the spot asset on or before a predetermined date (maturity date) at a certain price. A put option gives the holder the right, but not the obligation, to sell the spot asset on or before a predetermined date (maturity date) at a certain price (strike price). Such contracts include:

- calendar year base-load strip options
- financial year base-load strip options
- average rate base-load quarterly options.

Whenever I have analysed using futures and options to give price certainty I have found them to be very expensive; that is, it's the same as the windscreen insurance analogy. It does provide certainty over some or all of your load though, if you want that certainty at the expense of possible savings. In reality, if that's what you want you may well be better off locking in a short-term fixed retail contract.

Over-the-counter products

Over-the-counter (OTC) products are similar to the exchange-traded products except that they are a direct transaction with a market participant or intermediary for an electricity product that financially settles against the electricity market settled price. There is a higher counterparty default risk as the transactions are not novated by a market clearing house. The parties need to make their own assessment of the other party's credit status and risk.

Let's have a look at the major over-the-counter products.

Swaps are essentially a contract to swap the difference between a fixed and floating rate. With a straightforward vanilla swap, one party contracts with another to exchange a floating electricity price for a fixed price over a defined volume for a specific period in time. For example, an end user would enter into a contract with a swap provider to pay a fixed price for a fixed block of electricity over a set period of a specific quarter. The average spot price over that quarter will be the settlement price, and the end user would pay (spot price is lower than fixed) or receive (spot price is higher than fixed) the difference between the swap fixed rate and the settlement price. Settlement may also occur at predetermined intervals throughout the contract period.

The hedge calculation is the same as that for an exchange-traded futures contract but the contract size can be set at an agreed, tailor-made volume.

Sculpted swaps are contracts to swap the difference between fixed and spot rates for different predetermined load volumes over predetermined time intervals. For example, a sculpted swap contract could be established where the end user pays or receives the difference between the spot and fixed prices for 13.5 MW between 7:00 am and 9:30 am and between 3:30 pm and 7:00 pm, and receives or pays the difference between the spot and fixed rate over 24 MW for all other time periods for the contract duration. Such a contract offers a lower cost swap than a vanilla swap for 24 MW.

Swaps are also known as contracts for difference (CFDs) or fixed-for-floating contracts.

Caps are also known as one-way CFDs, and provide protection for a buyer above a predetermined price for a predetermined volume and period of time. The hedge calculation is the same as that for an exchange-traded options contract but the contract size can be set at an agreed, tailor-made volume.

Floors are similar but opposite to caps in that they set a maximum floor price. Generators may use floors to guarantee a minimum price.

Collars are a combination of a cap and a floor in which a minimum and maximum price can be set. Collars allow buyers to decrease the cost of a cap by selling a floor.

Caps, floors and collars are usually settled at predetermined intervals throughout the contract.

Asian options are OTC options in which the settlement price is based on the average level of price throughout a predetermined period. They are also known as "fixed strike Asians". Asian options can be in the form of a call or a put, and are typically used as the basis for a cap (call) or floor (put).

The risks of using OTC derivatives to hedge risk are:

- *Basis risk:* once again the hedge is not a perfect hedge as the swap contract is settled on the time-weighted average price whereas the price paid for the physical is a load-weighted average price. OTC swaps have an advantage over futures contracts as the swaps can be tailor made to better match the load profile. The basis risk can again be mitigated with a load-shed management strategy.

- *Over-hedging:* again this risk is mitigated by tailoring the contract to better match the expected load profile and under-hedging the likely forecast electricity consumption.

- *Default risk:* this risk is higher than with exchange-traded derivatives as the contract is not novated; that is, there is counter-party risk of default. This risk is managed by ensuring the counter party has a strong credit rating or by securing bank guarantees.

Providers of OTC electricity financial instruments are:

- retailers – who can wrap the product up as a retail contract

- generators

- financial intermediaries such as major banks.

End users have generally been poor users of such electricity risk management products due to:

- the predominance of retail fixed-price contracts

- electricity purchasing carried out by supply departments without treasury involvement
- a lack of understanding of the level of risk exposure
- a lack of understanding of risk management products
- a lack of understanding of the value that financial risk products can deliver in combination with load curtailment
- the requirement for "hedge accounting" in the half-yearly and annual financial statements, including the requirement for mark-to-market data to account for the current value
- boards and senior management are wary of the use of derivatives after highly publicised cases involving the misuse of derivatives, such as Enron, Pasminco and NAB.

In my experience the cost of the hedge product usually more than negates the potential upside of a wholesale market strategy. If you wanted to hedge 100% of your load to eliminate the risk or fluctuations in monthly costs then you are just as well off with a retail contract and then you don't have to worry about the added complexities of financial risk management and hedge accounting.

If you wish to hedge just part of your load – for example, the load that you can't curtail – and leave the rest of your load exposed to the market then a hedge strategy may be appropriate.

Physical hedging

Many facilities have critical equipment electricity supply backups to ensure the integrity and safety of the equipment in the event of a power failure. Examples include:

- cement plants that have backup diesel generators to continue to rotate kilns in the event of a grid supply outage
- hospitals with backup generation to ensure the continued supply of electricity to critical life-saving equipment and lighting
- buildings with backup supply for lighting
- shops with backup supply for refrigeration.

It's very common to have backup generation in the event of main grid supply outages. Some businesses also have backup generation for high electricity spot market events. The backup generation effectively works as a cap: the diesel or gas generation can be set to turn on when the cost of this generation is below the spot market price. The variable cost of that generation is the cap price.

There are obviously high costs in setting up backup generation, including high costs with grid synchronisation requirements. There are also complications with large backup generation (greater than 30 MW) that is synchronised with the grid as the owner of the generator would need to become a registered participant and comply with market rules and directions.

Whenever I have researched setting up physical hedging with gas or diesel generation to manage the risk of high price spikes, I have found that the business case does not stack up as the price spikes above the break-even point for the generator are very infrequent. It can work, however, for businesses that already have the generation installed for purposes such as critical equipment backup supply, or they may have had the generators from another project that no longer requires them.

The cheapest form of a physical hedge is demand side management (DSM). Demand side management was examined in detail earlier in this chapter: from the perspective of an end user, it can be described as the management of electricity consumption to increase use during low-cost periods and decrease use during high-cost periods, including short-term price spikes. The term is also used to encompass electricity efficiency measures in response to increasing prices; that is, reducing electricity consumption during both high- and low-cost periods.

Load curtailment has the potential to provide significant financial benefits to demand-side participants and contribute to the overall efficiency of the energy market. By analysing electricity price patterns by time of day, day of week and month of year, businesses can schedule their activities to minimise their overall annual average price. For example, if a plant has an annual shutdown it could schedule that shutdown during the historically high-priced months. Plants that have weekly scheduled maintenance for individual assets could schedule them for a historically high price day, and time them to start just prior to the high-price morning period. For operations that run 24 hours a day, they could maximise the loads that are not required to run all the time during the historically low price periods – for example, overnight and weekends – and minimise their operations during the historically high price periods of early day time and late afternoon and early evening.

Hybrid retail contracts

Some retailers that are more flexible offer a hybrid contract whereby they provide a wholesale market pool price pass-through arrangement and then offer, effectively, the option of quarterly swaps for up to two years in advance for blocks of 1 MW or multiples of 1 MW. The customer can then choose what quarters they would like to hedge and for how much load.

The customer can make an assessment of the value of the swap price compared with recent wholesale market pricing, pricing trends and their view of future prices. For example, a large customer could set a policy that they must hedge 50% of their load at least one quarter out from the commencement of a market period, but could then have flexibility to hedge up to, say, 80% of their load depending on the attractiveness, or otherwise, of the swap prices being offered. This type of strategy is usually only available to large consumers in excess of 10 MW demand size, or 50 GWhr per year.

Another form of emerging hybrid contract is an interesting renewables-backed retail supply arrangement – usually large-scale solar – that offers the customer a long-term, attractive fixed electricity price during the period of solar electricity generation, and then the remainder of the electricity is supplied at the wholesale market pool price. This still exposes the customer to the higher priced periods in the very early morning and the early evening, but generally the period of solar generation, provided the sun is shining, is during the higher priced day-time period. The customer has greater pool price exposure during the relatively low priced night-time period.

The term of these contracts is usually between five and ten years. Some end users balk at locking in a contract for such a long period, but in the past such long-term contracts were the norm. I have reached agreement on a 15-year electricity supply contract that was the most competitive offer for year one of the contract and quickly became a contract that had pricing well below the retail market pricing. This agreement had no take-or-pay commitment (minimum volume requirements). Long-term contracts tend to favour the buyer, provided you are not outwitted by the annual price escalation clause.

CHAPTER 14

Step 7: Reduce your electricity consumption and improve efficiency

The cheapest electricity is the electricity that you do not use. This chapter examines some of the common electricity consumption waste that occurs, and how you can identify that waste and reduce or eliminate it to reduce business costs.

Australia has enjoyed very low electricity costs over many decades due to our abundance of energy sources such as coal and gas. State governments originally established large coal-fired power stations to provide households and industry with cheap and reliable electricity. This was a major source of Australia's competitive advantage. As a consequence, capital investment decisions were largely based on the capital cost of the appliance, equipment, building or plant, and not necessarily on energy efficiency.

Now that electricity costs are escalating there is great opportunity to reduce costs through energy-efficiency initiatives. Many businesses are implementing energy-efficiency programs that are delivering substantial reductions in energy use. However, many more businesses struggle to get traction in identifying and implementing such initiatives.

The major reasons for this are around people. Employees and management need to be fully engaged in energy cost reduction, and to do this they need to understand energy use in the business and have visibility of the data. At least one person on every site should have a primary accountability for energy spend.

Most businesses have a lack of visible energy information, a lack of measurement tools and a lack of understanding of how the different elements of electricity tariffs, electricity demand and electricity usage combine to drag down profits. As the old adage goes, "you can't manage what you don't measure". Ask yourself these questions:

- Do you have an individual in your business accountable for energy spend?
- Does your business visually monitor and track energy use and energy use patterns?
- Do your employees have visibility of the data and flag abnormal usage?

Many large businesses with a large energy spend employ an energy manager to "manage energy". However, the site operations manager, or equivalent, needs to be accountable for energy spend. They need to understand their energy usage and costs and measure their performance. Ideally, their personal performance should have an energy cost metric.

In this step I will show you how to drive down electricity consumption in your business.

Everyone can reduce their electricity consumption

Every household and business can reduce its electricity consumption. Most choose not to.

Energy savings at home

In the simplest example, think of your own home. There are many ways to reduce electricity consumption. Some require investment while others require minimal effort. Everyone knows that you can reduce household electricity consumption by not leaving equipment running that is not required, such as televisions, computers, stereos, air conditioning and heating. But most people don't pay much attention to switching appliances off.

We also know that we don't need to chill the air down to 18°C in summer or warm it to 25°C in winter. But most people do.

We also know that when we buy appliances we could choose them based on an energy rating and the amount of power that they draw. But many people don't.

We know that when we walk around the house at night with the lights off we can still see where we are going due to the many little standby lights that are on with all of our appliances, such as televisions and set-top boxes, computers, stereos and microwaves. We know that we can turn them off at the wall to save on electricity, but most people don't.

We could even get serious about reducing energy consumption by putting shades over windows to stop the heat radiating into the room in summer, stopping the draft under doors causing additional heating or cooling, not using clothes driers, not running the pool pump as long as we do, changing out fluorescent light bulbs with LED lights, or by replacing our inefficient washing machines, fridges, dishwashers and other appliances with more modern and more efficient items.

We don't do all of these things because individually they may not cost much money on a daily basis, it's a bit of a hassle, it's a bit difficult and inconvenient to switch appliances off at the wall, it's not the most important issue on our minds, or because the initiatives cost a bit of money upfront.

Energy savings in the business

It's like that in our businesses, except our employees aren't paying the bill out of their pocket so the level of attention and care is even lower. So how can we drive electricity consumption savings in the workplace if our employees aren't really even interested in doing it at home and saving their own money?

The answer is by focusing on it and making it a point of emphasis within the business. The best way to focus on electricity efficiency in the workplace is to engage employees in a team with a clear objective around energy efficiency. In Step 1 we formed an energy cost reduction team to develop the strategy. This team could be the same team that investigates efficiency opportunities, or new sub-teams could be formed, led by a member of the strategy team.

The energy-efficiency audit

The first stage of any energy-efficiency project is to conduct an audit of the current situation. Fortunately there are many resources on the internet that provide a guide to conducting energy-efficiency audits, and there is also an Australian/New Zealand Standard for Energy Audits AS/NZS 3598:2014.

You may also wish to engage an experienced energy-efficiency auditing consultant to carry out the audit for you. The advantage of this is that they already know the sorts of things they are looking for, and can carry out the audit usually more quickly and more thoroughly than if you had to do it with little starting knowledge. Most auditors will quite heavily utilise resources at the site so that there is very good engagement with employees.

So what are the opportunities that an audit will identify? In this chapter we're going to examine how you can find areas of wasteful energy usage in your business, and what you can do about it.

Process efficiency

In a processing facility most of the energy is used in the process, so small improvements in process efficiency can result in large energy savings.

A process runs most efficiently when it's running continuously with minimal changes and breakdowns. Quite often, operations people understand this and already do have a focus on steady operations and energy consumption per unit of product. However, this measure, while important, is usually subordinated to other important metrics and not given the focus it requires if energy savings are a business objective.

Let's take the example of cement.

The process of grinding cement clinker that is made in cement kilns into the cement powder uses between 50 kWh and 60 kWh per tonne of product. Let's say we are targeting a 10% improvement from 60 kWh per tonne, which is a realistic target. This is a reduction of 6 kWh. At an average spot price of, say, 7.5¢/kWh, this equates to a saving of 45¢ per tonne. This does not sound much, but when applied to an annual production of 2 million tonnes this equates to $0.9 million per year.

So should this metric have high visibility, with any aberrations thoroughly investigated to determine and eliminate the cause? Most certainly, if your business is serious about reducing energy costs.

Aside from stability, some of the other opportunities to improve process efficiency and reduce electricity costs are:

- insulation to reduce heat loss
- variable speed drives on fans to replace mechanical dampers
- variable speed drives on pumps, rather than recirculation or throttling valves
- correct sizing of process equipment
- proper maintenance of equipment (such as cleaning of heat exchangers, ensuring minimum pressure drops)
- proper operation of equipment (such as standardised operating parameters and visual metrics).

Utilities

While the main process uses the bulk of the electricity, what is commonly underestimated is the amount of energy used in site utilities such as compressed air, heating, cooling, lighting, and cooling towers. And it's in the utilities area where most of the losses or inefficiencies occur.

Compressed air systems

One of the biggest culprits is compressed air. In some businesses, compressed air can contribute up to 70% of total costs yet the inefficiencies can often be quite large. The US Department of Energy estimates that up to 30% of compressed air production is lost as leakage, and up to 50% is wasted through a combination of leakage, inappropriate use and artificial demand.

Compressed air needs to be viewed as a total system and not just the operation of the compressors. The electricity cost for running compressors is commonly greater than 70% of the total lifecycle costs of the compressor.

The **first step** in auditing compressed air systems is to map the entire system. This involves finding, or drawing up, a piping and instrumentation drawing of the entire compressed air system, and then listing all of the compressed air users and equipment on that system, including their location, physical layout, condition, piping and controls. This is *not* a desktop exercise. The equipment and their details need to be checked in the field. Often it's found in the audit that the actual system use can have significant differences to the design. For example, I have found air being tapped off to provide cooling for bearings or process hot spots.

The **second step** is to take measurements to determine the actual demand and compressed air system efficiency. The measurements you should take are power (kW), flow (cfm) and pressure (psi or kPa) over a period of one to two weeks. You want to be taking measurements that cover high-demand periods such as working days, and low-demand times such as night time and weekends. This data is very valuable as it gives you an understanding of total usage, usage patterns, efficiency and cost.

A rule of thumb for an efficient plant air system is that the power draw should be < 21 kW/100 cfm. A system that is running at more than 21 kW/100 cfm is burning cash. This could be due to compressor design, system design or compressor control, or all three.

We can also calculate the cost of the air as we know the kWh consumption and so can apply the electricity energy cost plus the network energy cost to calculate the compressor costs per hour, day or week. We can also determine the relative impact of compressor use on maximum demand and network costs. And we can calculate the efficiency of the compressed air system in kW/100 cfm to understand where the system rates with respect to the rule of thumb.

The **third step** is to analyse the system.

We can plot time-of-day and day-of-week flow plots to understand patterns of use and potentially identify waste. Is there a cycle that occurs that drives consumption very high for a short or regular period of time? The driver of this high use can be hunted down and possibly eliminated.

For plants that operate only during the day or on weekdays, what happens to consumption when the plant is not running or has minimal operations? Is there still an appreciable compressor load? This also needs to be investigated.

We can then take a closer look at the system components and characteristics. This involves identifying inappropriate uses of air, such as for drying, blowing or sparring. Each of these is an inefficient and expensive use of plant air. Low-pressure electric blowers should be used instead.

In process plants that operate at high temperatures where refractory is installed to protect the steelwork, when the refractory fails in patches it can produce a hot spot on the steel. Quite often, when this occurs, compressed air is used to blow on the hot spot to cool it down. This is very costly and inefficient. The level of cooling is dependent on air flow, not air pressure. Electric fans with or without mist sprays are much more effective for cooling and consume far less electricity.

Similarly, using compressed air for pneumatically driven devices is inefficient. The ratio of power used to produce the compressed air to the power required to actuate or drive a device is approximately 7:1. Electric power should be used instead.

Of course, we will also identify air leaks. This can be done through simple means such as listening for leaks or through the use of technology such as ultrasonic meters or thermography.

Common locations of leaks are:

- pipe joints and flanges
- couplings and fittings
- poor thread sealants
- filters, regulators and lubricators
- blowdown valves
- condensate traps
- poorly maintained pneumatic tools and equipment.

Compressed air systems are designed to operate at the pressure required by usually a small number of high-pressure users. For every additional 2 psi required in the system there is a 1% reduction in efficiency. It's far better to split low-pressure and high-pressure systems or have air boosters for high-pressure users.

It's very important to have the correct size and type of compressors supplying the plant air. It's best to have a larger centrifugal or modulating control rotary screw compressor supplying base-load air. These compressors should run the majority of the time as they lose efficiency at lower loads. A multistage VSD compressor should be used to swing on and off as they have better partial-load efficiency. Quit often you will find two or more large compressors with one unit coming on and off to maintain pressure.

This is very inefficient and will drive the kW/100 cfm very high. A master system controller should ensure that the compressor system operates at maximum efficiency and lowest kW/100 cfm.

The **fourth step** is to develop the recommendations, including the business case if capital is required.

The **fifth** and **sixth steps** are implementation and verification, and ongoing monitoring. It's critical that you verify the kW/100 cfm has dropped after implementing the recommendations, otherwise you will have achieved nothing. This can be the case – for example – if leaks are eliminated but the compressor master controller isn't tuned correctly to optimise efficiency.

Ongoing monitoring, similar to electricity monitoring, is required to ensure that the gains that you have made don't deteriorate with time back to where you were.

Heating, ventilation and air conditioning

The Australian Government estimates that heating, ventilation and air conditioning (HVAC) in non-residential buildings consume approximately 9% of all electricity generated in Australia, and account for 40% to 60% of electricity consumption in those buildings. Yet HVAC systems are generally not well managed and are often plagued with inefficiencies.

Issues that I have observed include:

- external inspection hatches that have "fallen off" on roofs, sucking in hot air on the cool-air side of a compressor
- hatches within ceiling spaces "falling off" and providing a chilled atmosphere in the ceiling space
- fans running when compressors have tripped
- dampers failing, causing some rooms to be chilled and adjacent rooms to be hot and uncomfortable
- lack of adequate refrigerant gas, causing systems to run flat out, drawing maximum power without meeting the objectives of the system.

Systems can quite easily be optimised to switch off when not required overnight and during weekends. You can set control bands around cooling temperature set points and heating temperature

set points, and minimise the amount of air that is ventilated and replaced with fresh air as far as practical. You can use outside air for cooling when outside temperatures are lower than inside. And it sounds dumb to say, but ensure that the system is not heating and cooling at the same time – it happens.

Often, HVAC systems do not receive adequate maintenance which can lead to a drop in efficiency of up to 40%. Check that filters are clean, refrigerant levels are adequate, chilled and heated pipework are correctly insulated, and that actuators and dampers are working correctly. Sensors also need to be checked for accuracy, and the control system should be checked to ensure that equipment is not bypassed or operating incorrectly.

Some systems use fixed-speed fans with dampers to control flow, meaning that the fans are running at maximum speed all of the time even when not required. By installing a variable speed drive (VSD), fans can be slowed down to meet demand without a damper – this will significantly reduce power draw.

Most buildings will have insulation in ceilings and walls, but for buildings that don't, electricity consumption can be reduced by 25% by installing insulation. Shading of windows can reduce the solar heat gain by 70%, and identifying and eliminating air draught can reduce heating requirements by up to 40% in cooler climates or seasons.

Computers and their screens, lights and photocopiers are, aside from people, the major sources of heat within office buildings. Switching them off when not required, installation of timers, and using sleep mode for computers and photocopiers can significantly reduce HVAC demand.

Cooling water towers are a key component of HVAC systems and many other industrial water cooling systems. It's important to ensure proper inspection and maintenance of cooling tower systems to minimise electricity consumption. Inspections should ensure that there is no equipment damage, air or water leaks, and that there is no fouling of equipment, including water distribution pipework. Conductivity probes need to be checked to ensure that there is not excessive blowdown of water. Proper chemical treatment of the water should be conducted by, usually, third-party specialists. Variable-speed drives on fans and pumps can be used to reduce motor electricity consumption compared with fixed-speed drives with dampers or valves, and the control system should be optimised to ensure the minimum number of required towers are running.

As with the other systems, proper measurement and monitoring should be implemented to ensure early detection of problems that cause excessive electricity use. Monitoring should include inlet and outlet water flows, water conductivity, and electricity demand and consumption.

Refrigeration and chilling systems

Refrigeration and chilling systems may be significant power users in some businesses.

How do you determine if your refrigeration systems are working efficiently? An energy balance can be estimated by determining the theoretical electricity requirement based on design specifications and the volume and products stored, and this can be compared with the actual electricity consumption. The difference is unaccounted for heat gain. This can be quite a detailed and technically challenging operation, but it is very important for a business that uses refrigeration as a major process.

The heat gain can come from things such as:

- inadequate or broken insulation on pipework
- air leakage through seals and airlocks
- too frequent access into the area by employees or customers
- pumps continuing to run even when the target temperature has been achieved
- the set point temperature being too low
- solar gain.

Some initiatives to reduce heat gain are:

- shading to reduce solar gain
- optimising the control system to ensure minimum equipment run time and flows to meet the system objectives
- raising the coolant temperature and the temperature set point if possible
- installing variable-speed drives to replace fixed-speed drives

- minimising access to the cool area by staff and customers if possible
- selecting a higher heat capacity secondary cooling fluid.

Regular maintenance of the system is required to ensure that all plant and equipment is running per design and nothing has been broken, fouled or bypassed.

Lighting

Lighting can account for up to 40% of electricity costs in commercial buildings and up to 7% of industrial energy use, and savings of 75% to 90% can be achieved through smart thinking and the use of energy-efficient lighting.

The simplest initiative is to turn lights off when not required. Think about your own industrial site, office building or home. Do you see lights left on during the daytime or in rooms or areas that are not being used? Simply encouraging a culture of turning lights off when not required and automating lights to start up and shut down at night and in the morning can save an appreciable amount of electricity. Installing skylights can reduce the requirement for indoor lighting during the day. And in combination with window glazing, reflective film or shade, windows can be a good source of light in offices during the day.

Similar to other energy-efficiency initiatives, maintenance is important to ensure electricity consumption is kept at a minimum. Over time, many lamps produce less light but draw the same power. Lamps, reflectors and diffusers accumulate dirt, and this reduces light output. Sometimes additional lighting is added as a response, rather than cleaning the fittings.

A lighting survey as part of an energy-efficiency audit can identify areas where the lighting is above standard so that the number of lamps could be reduced. A good lighting survey can also provide advice on whether lighting retrofits and/or upgrades are financially viable.

Upgrading lighting systems, while sounding simple, can be quite complex. There are many retrofit and upgrade options beyond just changing out all lighting for LED lighting. While efficient lighting systems such as LED systems do significantly reduce energy consumption and have much longer life than older technology, the capital costs in such a change may be quite high. There may also be better options than LED systems.

Payback periods can vary considerably. Depending on the capital cost inclusive of installation, payback periods can vary from one to five years, or longer. The quality of the lighting used is critical in achieving the target payback. Typically, you should expect a five-year guarantee from a supplier of an LED system.

Another issue is that LED lighting systems can fail in environments where there is a high level of vibration. Trialling different options can give you confidence that a particular lighting system will meet the desired performance level in difficult environments.

Some businesses choose to bite the bullet and change out their entire lighting for LED and/or other systems. This may be due to a state or Commonwealth grant scheme to subsidise the cost of installation, or for cost-efficiency reasons in getting a contractor in – with the necessary equipment – for a campaign of change outs.

Other businesses choose to replace or retrofit systems as they reach end of life. The problem with this process is that the total cost of replacement installation over the course of several years may be much higher due to inefficiencies in labour and equipment. I once found a full-time, permanent electrician on a cherry picker changing out an external light for a new LED system. He was using a hired cherry picker, and so had another full-time, permanent electrician as his safety observer. It took two electricians and a cherry picker to change a light fitting. I'm not sure whether they were scheduled to change other light fittings that day, but it was an expensive way of doing it.

Changing lighting systems based on end of life can be an expensive exercise in aggregate over several years. Choosing whether to retrofit, upgrade, replace all at once or based on end of life will depend on the objective set during the strategy development stage.

* * *

The cheapest electricity is the electricity that you do not have to consume. Many efficiencies can be achieved with zero or little cost, while some require a larger capital investment. The efficiency targets that you set will be based on your business's energy objective and the priorities from your energy audit.

CHAPTER 15

Step 8: Understand renewable electricity

Free power

What if I told you that you could get your electricity for free in the future? All you had to do was pay a fee upfront and then you would get free electricity for life. Does that sound too good to be true? Well, it isn't.

Almost everyone, whether at home or in business, likes the idea of installing solar power generation. Who would *not* like free electricity from the sun after covering the initial purchase and installation costs?

Renewable electricity generation is growing exponentially worldwide. Even though there are many people who have a philosophical problem with using taxpayers' money to subsidise renewable generation, many don't seem to have a philosophical problem with states using taxpayers' money to prop up or keep open uneconomic coal-fired generation. Irrespective of these different views, renewable generation is being built by the smart fast-movers who are ahead of the game.

A good example is Sun Metals, in Townsville, who are investing an estimated $200m-plus in a 116 MW solar farm. Is renewable generation core business for them? No, but making zinc at the lowest possible cost to maximise returns to their owners is.

The problem in the past has been that it has taken a long time to generate enough electricity to cover the initial costs. High subsidised feed-in tariff rates for domestic solar have certainly made investment in household solar attractive, but businesses do not get the subsidised rate, and instead receive a rate that is more closely aligned with the market rate. In many cases the business will consume all of the electricity that it generates itself.

Renewable energy certificates

The benefit that a business does get in investing in renewable energy is revenue from generating renewable energy certificates. There are two Commonwealth Government renewable energy schemes: the Large Scale Renewable Energy Scheme and the Small Scale Renewable Energy Scheme.

The Commonwealth Government had an aim of achieving 20% renewables generation by 2020. The renewable energy target was modified in 2015 to achieve 33,000 GWh of large-scale renewable energy generation by 2020. In effect, this means that by 2020 the amount of renewable electricity generation as a proportion of total generation will be approximately 23.5%.

To achieve the national target, the government determines how much electricity is required to be produced by renewable generation each year. In 2017, the large-scale renewable energy target was 14.22%. This means that 14.22% of the electricity that your business uses is required to be backed by large-scale generation certificates (LGCs).

The Small-scale Renewable Energy Scheme operates a little differently – it was designed to encourage households and small businesses to invest in renewable energy and achieve a financial return. Small-scale schemes (< 100 kW solar and < 10 kW wind) generate small-scale technology certificates (STCs). Each scheme generates all of the STCs for a deemed period upfront in the first year of installation. The deeming period prior to 2017 was 15 years, but from 2017 this period reduces by one year, every year, until the planned end of the scheme in 2030. So the longer you wait to invest in a small-scale scheme, the fewer STCs you will receive.

At the time of writing this book, Australia's investment in large-scale renewable generation has not met its longer term obligation so the price of certificates has risen to approximately $85/MWh. This is important to large electricity users as they have to pay this price on top of their retail or wholesale price for a proportion of their electricity use, and that cost may be very high.

There are many types of renewable electricity generation in various stages of development and economic viability. These include the more traditional rooftop solar, ground-mounted solar and wind generation, established but more expensive technology such as geothermal, hydroelectric, wave energy and bioenergy, and long-established energy or heat recovery systems.

Renewable energy may, or may not, become part of your electricity cost reduction strategy. For many businesses, investment in renewable energy has been a very important component of their energy strategy to reduce electricity costs immediately and to protect them against future price uncertainty. More importantly, it has allowed them to take control of their own energy future.

With renewable energy there is not a standard technology or approach that suits every business. There is often a specific technology that best suits the unique circumstances of each business.

A business with a large amount of roof space may be well suited to roof-mounted solar. Businesses outside of cities with an abundance of land may well be suited to ground-mounted solar or even wind generation if they are in high-wind regions. A mining company located in a remote area close to a near-surface geothermal resource may choose to invest in geothermal. A process that requires or generates large amounts of heat may be well suited to heat recovery, while a process that is in an agricultural area or is associated with agriculture or forestry may be well suited to electricity generated from bioenergy. Another company with the capacity to store water at different locations and heights could be well suited to investing in hydroelectric technology.

The point is that every business is unique and has its own unique opportunities to employ renewable electricity generation technology that suit its industry, location, process technology, electricity usage profile, electricity usage intensity and balance sheet.

Solar generation

Solar power is the most viable renewable option for businesses. The cost of household solar and commercial-scale large solar is plummeting. A typical cost for a ground-mounted solar farm is about $2.00/Watt, or $2m/MW. A roof-mounted system is half the price at $1.00/Watt. An efficient, ground-mounted solar farm should be able to average about 4.2 MWh per day per MW of installed capacity over the course of a year. Of course, geography plays a large part in this. The further north you go in Australia the higher the solar quality, and if you can avoid the tropical north then the higher the efficiency.

A larger solar generation example

Let's look at a very simplified example of the economics of a businesses investing in their own modest size, large-scale solar project. We will look at a 250 kW roof-mounted solar generation project in South Australia, installed at a cost of $1.00/Watt. This should generate, on average, 4.2 MWh/day and 383.25 MWh over a full year.

The equivalent value of the electricity generated is $41,805 per year, based on 2017 wholesale market daytime prices of $109.08/MWh. An investment of $250,000 will produce $41,805 per year of electricity value for a minimum of 20 years, with those assumptions.

But it gets better.

As you are generating renewable energy, you are eligible to create an equivalent amount of large-scale generation certificates. These can be surrendered to meet your own obligations, sold on the market, or sold direct to a retailer or end user. The market price goes up and down depending on demand.

The government sets annual targets for how much renewable electricity should be generated. This then determines how much renewable energy is required to be purchased by consumers via their retailers each year. There is a lag in the investment in renewables, resulting in less renewable generation and so a shortage in renewable energy certificates. The high demand and lack of supply has pushed the traded price of renewables up to $85/MWh, and this does not look like dropping very quickly in the

next few years, but it will drop as new projects come online and the national target is met.

However, there is also uncertainty about what happens at the end of the current scheme in 2030, and many large-scale projects are holding off a final investment decision until they have more certainty. For example, if a project does not come online until 2022 then they will only be certain of nine years of LGC revenue. This could provide a floor under the LGC prices.

In our example, we will base our return over a 10-year life and an LGC price that starts at $80/MWh then trails off to $50/MWh by the end of the 10 years. This means that on top of the notional $109.08/MWh that we are earning from electricity generation, we are also earning revenue from renewable energy certificates.

The net present value (NPV) and internal rate of return (IRR) calculation are shown in the following table for this simplified example.

In this case, the IRR is 17% after tax and a payback period of 5.5 years. This is not a bad return but it is quite a long payback period.

The issue that most businesses face with such an investment decision is that an investment of this size crowds other investment opportunities. The simple payback on this project is five-and-a-half years. In a large publicly owned business there are usually a large number of capital investment projects being put up for approval in competition with other projects for a limited capital budget. A project with a simple payback period of more than four years is unlikely to get the business very excited. Often businesses mandate that cost-reduction capital projects require a two-year payback but that strategic projects such as an investment in new equipment to increase production significantly may have a longer payback period.

This book is about treating energy spend strategically, as with large energy users it is one of the largest input costs and one of the largest profitability levers. To achieve up to a 50% decrease in electricity costs you need to think strategically about electricity and make investments for longer term sustainable cost savings. If electricity costs are a strategic issue for your business then electricity cost-reduction projects should be measured by their rate of return rather than the blunt instrument of simple payback.

Cost$/W	$1.00
Size (MW)	0.25
Average generation/W installed	4.2
Average Daytime Spot price$/MWh	109.08
WACC	8%
Tax Rate	30%
NPV	$152,800
IRR	17%

Year	0	1	2	3	4	5	6	7	8	9
Costs	$(250,000)									
Electricity Generation		383	383	383	383	383	383	383	383	383
Electricity Value		$41,805	$41,805	$41,805	$41,805	$41,805	$41,805	$41,805	$41,805	$41,805
Forecast LREC Price$/MWh		$80	$70	$65	$60	$60	$60	$60	$55	$50
LRET Certificate Revenue		$30,660	$26,828	$24,911	$22,995	$22,995	$22,995	$22,995	$21,079	$19,163
Maintenance, cleaning & monitoring costs		$(5,000)	$(5,000)	$(5,000)	$(5,000)	$(5,000)	$(5,000)	$(5,000)	$(5,000)	$(5,000)
EBITDA impact	$(250,000)	$67,465	$63,632	$61,716	$59,800	$59,800	$59,800	$59,800	$57,884	$55,967
Depreciation (non cash)		$(12,500)	$(12,500)	$(12,500)	$(12,500)	$(12,500)	$(12,500)	$(12,500)	$(12,500)	$(12,500)
EBIT Impact		$54,965	$51,132	$49,216	$47,300	$47,300	$47,300	$47,300	$45,384	$43,467
Tax @ 30%		$(16,489.47)	$(15,339.72)	$(14,764.85)	$(14,189.97)	$(14,189.97)	$(14,189.97)	$(14,189.97)	$(13,615.10)	$(13,040.22)
NPAT Impact	$(250,000)	$50,975	$48,293	$46,951	$45,610	$45,610	$45,610	$45,610	$44,269	$42,927

Year	10	11	12	13	14	15	16	17	18	19	20
Costs											
Electricity Generation	383	383	383	383	383	383	383	383	383	383	383
Electricity Value	$41,805	$41,805	$41,805	$41,805	$41,805	$41,805	$41,805	$41,805	$41,805	$41,805	$41,805
Forecast LREC Price $/MWh	$50	$–	$–	$–	$–	$–	$–	$–	$–	$–	$–
LRET Certificate Revenue	$19,163	$–	$–	$–	$–	$–	$–	$–	$–	$–	$–
Maintenance, cleaning & monitoring costs	$(5,000)	$(5,000)	$(5,000)	$(5,000)	$(5,000)	$(5,000)	$(5,000)	$(5,000)	$(5,000)	$(5,000)	$(5,000)
EBITDA impact	$55,967	$36,805	$36,805	$36,805	$36,805	$36,805	$36,805	$36,805	$36,805	$36,805	$36,805
Depreciation (non cash)	$(12,500)	$(12,500)	$(12,500)	$(12,500)	$(12,500)	$(12,500)	$(12,500)	$(12,500)	$(12,500)	$(12,500)	$(12,500)
EBIT Impact	$43,467	$24,305	$24,305	$24,305	$24,305	$24,305	$24,305	$24,305	$24,305	$24,305	$24,305
Tax @ 30%	$(13,040.22)	$(7,291.47)	$(7,291.47)	$(7,291.47)	$(7,291.47)	$(7,291.47)	$(7,291.47)	$(7,291.47)	$(7,291.47)	$(7,291.47)	$(7,291.47)
NPAT Impact	$42,927	$29,513	$29,513	$29,513	$29,513	$29,513	$29,513	$29,513	$29,513	$29,513	$29,513

A smaller solar generation example

Let's have a look at an example for a smaller business. For smaller businesses, a small-scale roof-mounted solar system may be more appropriate. In this case, the equation is a little different as the initial capital cost is much lower at $1.00/W, and the renewable energy certificates are different in that they are classified as small-scale technology certificates. These are paid upfront for up to 15 years but have a lower value determined by a calculation made by the government in order to fund these subsidies. Businesses can use this upfront revenue to offset the initial capital investment.

We will assume that this business runs 24 × 7 and consumes all of the solar electricity that is generated. This will not necessarily be the case, and, in reality, the actual load profile will need to be overlaid with the solar-generation profile, and exported electricity may not receive the market price. However, for simplicity we will assume all of the electricity is consumed by the business.

In this case the rate of return increases to a massive 50%.

The costs of solar have dropped sharply over the last five years and the competition to install projects has increased. Solar electricity generation is a financially favourable method of reducing electricity costs and meeting a strategic objective of a 50% reduction in costs. Any business that is serious about reducing electricity costs should look at the business case for solar.

Cost $/W	$1.00
Size (MW)	0.1
Average generation/W installed	4.2
Average Daytime Spot price $/MWh	109.08
WACC	8%
Tax Rate	30%
NPV	$76,825
IRR	50%

Year	0	1	2	3	4	5	6	7	8	9
Costs	$(100,000)									
Electricity Generation		153	153	153	153	153	153	153	153	153
Electricity Value		$16,722	$16,722	$16,722	$16,722	$16,722	$16,722	$16,722	$16,722	$16,722
Forecast SRES Price $/MWh	$35.00	$–	$–	$–	$–	$–	$–	$–	$–	$–
SRES Certificate Revenue (15-years upfront)	$80,483	$–	$–	$–	$–	$–	$–	$–	$–	$–
Maintenance, cleaning & monitoring costs		$(5,000)	$(5,000)	$(5,000)	$(5,000)	$(5,000)	$(5,000)	$(5,000)	$(5,000)	$(5,000)
EBITDA impact	$(19,518)	$11,875	$11,875	$11,875	$11,875	$11,875	$11,875	$11,875	$11,875	$11,875
Depreciation (non cash)		$(5,000)	$(5,000)	$(5,000)	$(5,000)	$(5,000)	$(5,000)	$(5,000)	$(5,000)	$(5,000)
EBIT Impact		$6,875	$6,875	$6,875	$6,875	$6,875	$6,875	$6,875	$6,875	$6,875
Tax @ 30%		$(2,062.58)	$(2,062.58)	$(2,062.58)	$(2,062.58)	$(2,062.58)	$(2,062.58)	$(2,062.58)	$(2,062.58)	$(2,062.58)
NPAT Impact	$(19,518)	$9,813	$9,813	$9,813	$9,813	$9,813	$9,813	$9,813	$9,813	$9,813

Year	10	11	12	13	14	15	16	17	18	19	20
Costs											
Electricity Generation	153	153	153	153	153	153	153	153	153	153	153
Electricity Value	$16,722	$16,722	$16,722	$16,722	$16,722	$16,722	$16,722	$16,722	$16,722	$16,722	$16,722
Forecast SRES Price $/MWh	$–	$–	$–	$–	$–	$–	$–	$–	$–	$–	$–
SRES Certificate Revenue (15-years upfront)	$–	$–	$–	$–	$–	$–	$–	$–	$–	$–	$–
Maintenance, cleaning & monitoring costs	$(5,000)	$(5,000)	$(5,000)	$(5,000)	$(5,000)	$(5,000)	$(5,000)	$(5,000)	$(5,000)	$(5,000)	$(5,000)
EBITDA impact	$11,875	$11,875	$11,875	$11,875	$11,875	$11,875	$11,875	$11,875	$11,875	$11,875	$11,875
Depreciation (non cash)	$(5,000)	$(5,000)	$(5,000)	$(5,000)	$(5,000)	$(5,000)	$(5,000)	$(5,000)	$(5,000)	$(5,000)	$(5,000)
EBIT Impact	$6,875	$6,875	$6,875	$6,875	$6,875	$6,875	$6,875	$6,875	$6,875	$6,875	$6,875
Tax @ 30%	$(2,062.58)	$(2,062.58)	$(2,062.58)	$(2,062.58)	$(2,062.58)	$(2,062.58)	$(2,062.58)	$(2,062.58)	$(2,062.58)	$(2,062.58)	$(2,062.58)
NPAT Impact	$9,813	$9,813	$9,813	$9,813	$9,813	$9,813	$9,813	$9,813	$9,813	$9,813	$9,813

Wind generation

Obviously, solar isn't the only renewable energy option. Another large-scale option is wind generation. This is much more technically and socially challenging to implement than solar, and requires more ongoing, higher cost maintenance than solar. Nonetheless, it's a very real option for very large electricity consumers.

Wind generation is more economically viable on the basis of capital cost per expected generation output than solar. Some of the challenges of this option are the large amount of land required for the turbines, potential environmental and perceived health effects, and the engineering and construction effort required compared with roof- or ground-mounted solar.

Another way to get exposure to the benefits of wind generation is to invest in a large-scale wind generation project. There are many ways of doing this, ranging from direct investment to contracting for long-term offtake agreements.

There are many proponents of large-scale wind generation projects, but in order for the proponents to gain funding for the projects, the providers of the funding most often require long-term offtake agreements. It is therefore possible to enter into a long-term contract for a potential project and lock in prices that are competitive compared with current prices for a long-term period of seven to ten years.

Many large electricity users have already locked in long-term pricing with large-scale renewable projects. These deals are structured in may different ways that may, or may not, provide the end user with the large-scale renewable energy certificates associated with the electricity purchased. They may also provide a future hedge against a price on carbon, if that is included in the offtake agreement. This is a critical consideration when analysing a potential offtake agreement.

An offtake agreement structure that may be attractive is one that provides the end user with a fixed unit price of electricity from the wind generation and the remainder of the end user requirements being supplied at the wholesale spot market price.

The structures for these offtake agreements can be totally flexible to meet the niche requirements of the end user.

Another possibility is for an end user to have a combination of their own large-scale solar, own or offtake agreement wind generation, and the balance of their requirements met by the wholesale spot market. Having both wind and solar helps to partially hedge when the sun is not shining or the wind not blowing.

Heat recovery

Another alternative energy option for many process industry businesses is heat recovery. Many operations produce heat as a byproduct of the process. Sometimes this heat is reused in the process, but often it's simply wasted into the atmosphere. There are now many options for commercially viable heat recovery systems that can generate electricity. This generation also qualifies for renewable energy certificates that help make the project more viable.

These projects may have a high capital cost and an extended payback period, but they also quite often have a very high internal rate of return. If rising electricity costs are a strategic issue and a threat to future business profitability then the capital investment needs to be considered on a long-term rate of return basis rather than a simple payback.

* * *

Whether you philosophically agree with it or not, the renewable energy revolution is occurring already, and it's happening very quickly. The smart and more agile businesses recognise this and are already investing directly in their own renewable energy generation to reduce costs, or are entering into arrangements with large-scale renewable energy projects. Does your business want to pay the exorbitant cost of $85/MWh for renewable electricity certificates, or receive that exorbitant amount?

CHAPTER 16

Step 9: Monitor, report and continuously improve

It's very important to measure and understand your electricity usage and costs if you are serious about reducing costs. If you do not measure it you can't tell if you are improving your electricity cost efficiency. You can't manage what you don't measure.

Why is monitoring so important?

As discussed earlier in the book, the electricity procurement process in many organisations is a three-year event involving the business going to tender to ask for electricity supply offers or paying a consultant to run the process. The offers come in, there is a bit of argy bargy conducted with the lowest price offers, and a contract is agreed upon and signed.

What happens after that?

Aside from the responsible energy manager having the occasional business lunch or dinner with the supplier and going to the odd footy and cricket match, the usual process is that the electricity is supplied and then billed on a monthly basis. The accounts payable person receives the invoice and sends it to the responsible business unit manager, who gives it a cursory glance and approves it along with the 30 other expenses he or she has to approve that day. It then gets paid.

Some businesses that engage consultants to manage their energy also receive monthly reports showing their daily consumption, maximum demand for the month, and other data. The manager may also have a slightly more detailed look at this report before deciding to address the important plant issues for the day.

What normally *doesn't* happen is a deep dive into the data to search for any incidents that have offset the cost savings, or any nuggets that lead to further cost savings. For example, a client of mine started to experience a step increase in maximum demand without any new equipment being installed and no change to the process. After this was detected in the load monitoring and investigation, it was found that there were some substantial compressed air leaks, and the second of two large air compressors was being run to meet compressed air demand. This caused an increase in the maximum demand charges by an annualised amount of $10,000. In this case, because it was detected early we were able to provide evidence to the network provider that the problem had been fixed and that the demand had dropped back to its previous level.

What should I be monitoring?

There are many parameters that you could and should monitor. These include:

- electricity consumption in kWh or MWh by site, area, process and/or device
- electricity load in kW or MW
- electricity demand in kVA or MVA
- power factor
- efficiency in kWh/unit of production
- electricity energy cost in ¢/kWh or $/MWh
- network costs in $/month
- metering charges in $/month
- renewable generation produced in kWh or MWh
- renewable energy value in ¢/kWh or $/MWh.

Load monitoring

Measurement of electricity consumption can tell you whether usage is increasing, decreasing or staying flat. It can flag changes in consumption patterns so that they can be investigated. If patterns of higher demand are discovered in the data then the source of that higher demand can be identified and potentially eliminated. Then new patterns can be identified and eliminated. Through monitoring and analysis you can continually strive to reduce demand charges.

If a business has adopted a pool pass-through and demand response strategy then it's important to monitor the half-hour load and verify that the load curtailment procedure is being followed diligently. If it's not then the operations manager needs to be alerted. This monitoring and analysis also gives feedback on how the average price paid is tracking compared with the spot price and the retail price that the customer would have paid with a vanilla contract.

Electricity load monitoring can likewise identify trends. Has the load stepped up or down? Why? Are there hourly, daily or seasonal patterns?

Electricity demand needs to be monitored closely at all times. Network charges are largely based on maximum demand. If your demand goes up above the contracted or agreed demand level, so do your network charges – in some jurisdictions, potentially forever.

In SA, if you negotiate to reduce your demand charges because of demonstrated changes in your operation and electricity demand patterns and then at some point within the subsequent 12 months the demand spikes back up again, you will be back-charged that amount to the date that you reduced it. So, it's very important that systems are put in place for immediate detection and control of demand spikes.

In other states, the impact is dependent on who the regional distribution network owner is. In some regions of Victoria, that one spike gets rolled into a 12-month rolling average of maximum demand and so the impact is not as pronounced as it is in SA. In other regions, the rolling maximum demand in the two preceding months determines the maximum demand charge. This means that the spike has a two-month impact on charges, while in some other regions the spike will only affect that actual month.

Power factor monitoring

In a similar way, power factor should be closely monitored at all times. If your power factor is low then you would be able to reduce your network charges by installing power factor correction. If you have already installed power factor correction and your capacitors are switched off or fail for some reason and you are at a high load then you could undo the savings you achieved with power factor correction, wasting the original capital and potentially having to pay back savings to the network provider.

Many businesses install power factor correction equipment to improve their power factor and reduce their maximum demand changes but then experience failures or deterioration in the power factor correction equipment without even realising it. As previously discussed, one client of mine had installed power factor correction in SA where the tariff reflects the maximum all time kVA unless negotiated down. Their power factor correction equipment was failing every now and again without the engineers realising. The electricians knew about it but didn't understand the impact on network demand charges. This site's network demand tariff increased by about $50,000 per year because of these infrequent failures and no-one noticed until we reviewed it. The accounts payable person had no idea and just processed the invoice, and the responsible operations manager just "rubber stamped" payment of the invoice each month. Constant monitoring and analysis would have detected this costly problem immediately, and corrective action could have been taken to prevent the increase in charges.

Other things to monitor

The list of things you could monitor to do with your power strategy is almost endless, but let's have a look at some of the most important things to keep an eye on:

- A true measure of efficiency is **how much electricity you have consumed per unit of product**. This ultimately tells you whether your efficiency efforts are resulting in improvements and savings.
- **Actual electricity cost per kWh** tells you how well you have managed your load to reduce costs. Whether you are spot exposed or have a fixed retail contract with peak and off-peak pricing, you can reduce electricity costs by planning production

to maximise exposure to low-cost periods and minimise exposure to high-cost periods.

- You should be keeping an eye on your **monthly network costs**, including breaking them down into their sub-components. If you track them on a spreadsheet you should be able to identify any adverse trends, errors and potential opportunities. This should include metering charges.
- If you have installed renewable or other generation then **how much you produce** and the **value of that production** is important to monitor so that you can maximise the financial benefit of your investment.
- Ideally the load, demand and power factor information is **monitored in real time** from a control room, with alarms at appropriate levels.
- **Daily reports** should be produced for staff to review the effectiveness of electricity management for the preceding day.
- **Monthly reports** allow managers to review how effective the electricity strategy has been, and should incorporate actual billing data.

With the right metering agent you can get almost live data that can be made visible to the appropriate staff. There are various software options that help to capture and analyse this data, however it can also be simply done using a spreadsheet.

If you are aiming to reduce your electricity spend then measuring and analysing electricity use and usage patterns is critical to managing that spend.

How to monitor your energy costs

So how would a business go about putting in place a monitoring and reporting system when their core product or service is the business focus?

Using an energy adviser

In larger businesses with high energy use, often, but not always, an energy manager is employed to manage the energy spend and develop and implement strategies to reduce costs. They may use some or all of the tactics that we have discussed in this book already, and they will likely employ many other tactics that are unique to their own business profile.

However, many businesses find it hard to justify employing a specialist energy manager. This is when using an energy adviser can add significant value, as you get the benefit of specialist energy knowledge for far less than the salary of an energy manager. Another advantage of an energy advisory service is that you do not run risk of an employed energy specialist building more and more knowledge, becoming more and more valuable to the organisation, and then leaving as they become even more valuable to another organisation.

The disadvantage is that you are outsourcing what may be a key strategic function and may then become reliant on the energy advisory consultant or business. To counter this disadvantage, you need to ensure that the adviser is keeping you up to date with what is happening in the energy markets, and that you are receiving detailed reports, at least monthly, of your energy usage patterns, anomalies and opportunities to continually reduce costs. The energy adviser should be educating you well enough such that you have the confidence that they are adding at least the value, if not multiples of, a full-time salaried energy manager.

Another disadvantage of an external energy adviser is that they may be providing advice to several clients, aside from your business. They may not be able to provide sufficiently bespoke analysis of your business energy usage to identify opportunities as discussed in this book as they have a large number of clients and a limited number of analysts. To counter this disadvantage it is important to ensure that the advice you are receiving is bespoke for your business and not vanilla reporting that just tells you how much electricity you have used and what you have paid each month. Larger advisory businesses have the advantage of very experienced analysts from the supply side of the market, who have very deep market knowledge and the ability to conduct informed market analysis and forecasting of potential trends in future pricing. They have well-established contacts with retailers and the ability to extract competitive offers from those retailers.

The question you need to ask yourself is: has my energy adviser recommended or achieved cost reduction for my business, outside

of a competitive supply process? Or are they simply telling me what I have used and how much have I paid? If they are just telling you how much you have used and what you have spent then you are missing out on the opportunities we have discussed to significantly reduce your costs.

Regular reporting

A business that is an active wholesale pool market participant and which curtails load in accordance with a load curtailment schedule may choose to require a daily report of electricity usage versus price to ensure that the load curtailment strategy is being effectively managed each day. This report can identify compliance, or non-compliance, to established procedures for managing load to reduce exposure to pool price volatility.

Other businesses that are exposed to pool pricing but have not instigated a load curtailment process due to their capacity to absorb the higher price spikes and their need to maintain a high level of operational availability to meet customer demand may be satisfied with a monthly analysis and report that shows how well, or not, the wholesale pool price exposure strategy is performing relative to a vanilla fixed-retail-pricing strategy and what anomalies or opportunities were evident during the month.

Measuring and managing equipment usage

Load monitoring can be done at the site level based on supply meter half-hour usage data, or it can also be done at the internal sub-board level at an increased granularity. The advantage of the sub-board level is that it provides even more detailed information on what equipment is being used and when. This can reveal information about load being drawn by equipment that is not really required. For example, are conveyors running while they are empty and not required? I have come across examples where crushers were still running when not required.

Many businesses say that they do not have the ability to curtail load in the event of high prices, but when a detailed study is performed at the sub-board level we often find that there is a large amount of equipment that's running flat out during high price events and then the load is reduced or turned off when prices settle back down. There is always an opportunity to curtail load.

Large businesses often already have process control systems and sub-board load measurement that enables them to monitor and manage load, in real time, and automatically load shed non-critical equipment during high price events. Many other businesses only have local control and little ability to measure and manage equipment usage in the context of reducing costs.

When we conduct an energy audit we identify equipment that can be curtailed due to high price events, and recommend measurement equipment and software that will allow those loads to be curtailed because of high prices.

We have clients that are 100% exposed to wholesale pool prices and have elected not to curtail as their equipment is at capacity to meet their customer demand. We know 2017 was a horror year for wholesale market pool pricing due to the closure of the Hazelwood coal-fired station in Victoria. The shock of this slightly earlier than expected closure flowed through to all the other states, including SA. Our SA clients with wholesale market exposure were still able to achieve a modest saving compared with the best fixed price retail offers that were made in late 2016. They certainly knew the risks when they adopted this strategy and gritted their teeth through the months of February to July when average wholesale market prices were high, particularly in February. But they then enjoyed the benefits of the sharply reduced prices in the second half of the year. The key was that they closely reviewed the monthly analysis to reassure themselves that they had adopted the correct strategy over the longer term. They were still happy not to employ a load curtailment strategy, but were aware of the potential savings should they choose to.

Monitoring prices

Our clients, and many businesses that have partial or full price exposure, use Global-Roam's NEM-Watch to monitor current and forecast prices in the market and to set up alerts at different price thresholds. NEM-Watch also enables them to gain an insight into how the market operates by showing:

- demand trends
- price trends
- interconnector flows

- what prices are doing in different regions
- what the generation fuel mix is in each region, including wind, solar and battery.

They also use the Global-Roam NEM-Review software to view specific market data sets to gain a deeper understanding of the drivers of the market. I use both extensively, and the data used for the generation of the charts in this book was sourced from NEM-Review.

Other end users who incorporate automatic demand response via their own process control software use the Global-Roam product deSide® to provide a feed of reliable live data.

* * *

Whether or not you use an external energy adviser, you should gradually educate yourself as to how the market works, what is driving it, and current trends in the context of the larger historical perspective. This then needs to be related to how it is impacting your business costs and profitability.

Regular, at least monthly, market updates and deep analysis of your business electricity usage, anomalies, patterns and trends are critical to ensure that you monitor, manage and lock in the cost savings we have discussed in the book.

PART IV

Getting down to business

CHAPTER 17

So, what now?

Energy costs are rising in Australia due to a number of factors that we have discussed in this book. But, you do not have to simply accept these rising costs as there are many things that can be done to substantially reduce them. You can take control of your own energy destiny to both protect your business from increases and boost profits.

If electricity costs are a concern to your business and are having an impact on profits then you should be treating electricity costs as a strategic issue. This means having a strategic plan and resourcing that plan to achieve optimum success.

By applying the 9-Step Electricity Cost Reduction Framework you can cost-effectively reduce your electricity costs by up to 50% versus a business-as-usual approach. Let's step through the process again:

1 You must develop a strategy that has a clear objective.

2 You should understand your own operational electricity usage profile.

3 You should understand how the market operates to identify opportunities around your profile.

4 You should understand your electricity bill so you can see the levers that you can pull to reduce costs.

5 You should seek to minimise network costs.

6 You should seek to reduce electrical energy unit costs.

7 You should strive for energy efficiency to reduce consumption.

8 You should look at renewable electricity supply options such as solar, wind and heat recovery.

9 You need to measure and monitor to find any problems, identify possible savings, lock in initiatives implemented, and make further improvements.

So, what now?

The first step towards reducing your electricity costs is to *take the first step*. Take action *today* to commence the development of an energy strategy. Book a time, location and people to attend. That will force you to think about and develop an agenda based on the framework that you have discovered here. Your business could be leaving tens of thousands, or even hundreds of thousands, of dollars on the table every year by making poor decisions about energy, and by just doing things the way they've always been done without realising there is another way. There isn't a day to waste.

You could do it yourself, you could delegate it to someone to lead on your behalf, or you could engage a trusted adviser to facilitate the process for you using the 9-Step Electricity Cost Reduction Framework. The most important thing is to do *something* now to initiate the process. (Altus Energy are, of course, willing and able to facilitate the process for you, and you can make contact with us through our website: http://altusenergy.com.au.)

I wish you all the best with developing and implementing your energy strategy.

Glossary

actual demand (kVA): a unit of measure of electrical apparent power which is the product of voltage (V) and current (A).

additional demand: the difference between the maximum demand measured in any half-hour period outside of the peak demand period and the agreed demand.

agreed demand: the level of demand in kVA that is used in South Australian network tariffs to charge the customer each month. It is the maximum demand measured in any half-hour period on workdays between 12 pm and 9 pm local time, during November to March. This time is known as the peak demand period.

bid stack: the stack of generator price bids to supply blocks of electricity in five-minute increments. The stack is arranged in increasing price order.

caps: an exchange-traded or OTC futures contract where the settlement price for the block of future electricity is the average market price adjusted such that no half-hour prices in the average exceed the cap value, usually $300/MWh.

contracts-for-difference (CFD): a swap.

demand (market): the electrical power (MW) requirements of consumers in a market region.

demand side management (DSM): managing or changing power consumed in response to market price signals or other financial incentives.

derivatives: financial contracts between two or more parties based on an underlying physical product (such as electricity).

futures: a type of derivative where the contract is a standardised contract traded on a public exchange and the settlement occurs at a fixed time in the future for a fixed price. The futures contract may be settled with delivery of a standard physical product (for example, wool) or financially settled with reference to a market price (for example, electricity).

gentailer: a single entity that has both generation and retail businesses.

half-hour wholesale price: the market region settled half-hour price that is the average of the six five-minute trading interval prices that make up the half-hour. Also known as the spot price or the pool price.

hedge: the use of derivatives or physical means such as diesel generation to manage price risk.

interconnectors: transmission lines that transport electricity between regional markets (states).

kilowatt-hours per tonne: a unit of measure of energy consumption per unit of product or energy efficiency.

kilowatt hour (kWh): a unit of measure of energy.

kilowatts (kW): a unit of measure of electrical power, also known as real power.

load curtailment: reducing load in response to a high price signal.

load profile: the power used by a customer each half-hour over a period of time, such as a week.

load shifting: moving scheduled load consumption to lower price periods.

long-run marginal cost: the total cost of generation of electricity per MWh for a particular generator including investment capital recovery.

mark-to-market: an accounting treatment to realise gains or losses on the value of a derivative before the derivative expiry.

megawatts (MW): a unit of measure of electrical power, equivalent to 1,000 kW.

off-peak: a deemed period of lower demand on weekends, public holidays, night time, and some daytime periods. The periods vary between networks, and network off-peak periods can be different to retail energy off-peak periods.

options: another form of derivative that gives the buyer of the option the right but not the obligation to purchase an underlying futures contract up to the settlement date of the futures contract.

out of the money: a term used to describe a contract price that is more expensive than the underlying physical market price ("in the money" or "fair value" are terms to use when the contract prices reflect the underlying physical market price).

peak: a deemed higher demand period on weekdays that attracts higher costs, typically day time and early evening. The periods vary between networks, and network peak periods can be different to retail energy peak periods.

petajoules (PJ): a unit of measure of energy. In this context, it is a measure of gas energy.

pool price pass-through arrangement: a contracted arrangement whereby the retailer passes through to the customer the half-hour wholesale or pool price for each half-hour rather than a fixed price.

power factor: the ratio of real power (kW) to apparent power (kVA).

scheduled demand: the electrical power (MW) requirements of consumers in a market region less non-scheduled generation (for example, small wind, solar, small local generation).

short-run marginal cost: the incremental cost of producing an additional MWh of electricity.

shoulder: a deemed period of moderately high demand between the peak period and off-peak periods. It does not apply in every region.

spot price: wholesale market or pool price.

swaps: a type of derivative that is not traded on a public exchange – it is an over-the-counter (OTC) contract. It is a contract between parties that allows the buyer to fix a price for a future block of electricity. The contract is settled financially at the end of the contract period with reference to the relevant electricity market settled price for that block period, usually a month or quarter.

take-or-pay commitments: a contractual obligation to consume a commodity such as electricity or pay for some or all of it even if it is not consumed.